THE PULSE OF THE PLANET

A STATE OF THE EARTH REPORT
FROM THE SMITHSONIAN INSTITUTION
CENTER FOR SHORT-LIVED PHENOMENA

THE PULSE OF THE PLANET

A STATE OF THE EARTH REPORT
FROM THE SMITHSONIAN INSTITUTION
CENTER FOR SHORT-LIVED PHENOMENA

Compiled and Edited by

JAMES CORNELL and JOHN SUROWIECKI

Harmony books

a division of **Crown Publishers, Inc.** 419 Park Avenue South New York, N. Y. 10016

Second Printing

© 1972 by Crown Publishers, Inc.

Library of Congress Card Catalog Number: 72-84309

ISBN: 0-517-500655
ISBN: 0-517-500663

Printed in the United States of America

Published simultaneously in Canada by General Publishing Company Limited

CONTENTS

ACKNOWLEDGMENTS

The success of the Center can be attributed only to the generous cooperation and assistance it has received from hundreds of scientists, research institutions, and volunteer correspondents throughout the world who have contributed daily event notification and information reports, photographs and specimens, and the preliminary results of field studies.

We especially want to thank the National Oceanic and Atmospheric Administration's Earthquake Information Center and the Seismological Network of the U.S.S.R. for the prompt reports of major earthquakes on both sides of the globe. Similarly, the U.S. Geological Survey and the Seismological Section of the Meteorological Agency, Tokyo, Japan, have supplied regular and excellent reports of volcanic activity. Numerous smaller seismic stations throughout the world have provided statistical data and reports of earthquake effects on local physical environments. Most important, scores of staff members at United States embassies throughout the world have supplied vital communications, information, and logistical support for scientists participating in field expeditions—usually under the most hectic emergency conditions.

If any individual deserves special thanks and credit, it must be Dr. Sidney Galler, the former Assistant Secretary for Science of the Smithsonian Institution, now of the Department of Commerce, who conceived and shaped the idea of a global scientific alert system.

Since its conception by Galler, the Center has found a home and unqualified support at the Smithsonian Astrophysical Observatory, thanks to Dr. Charles A. Lundquist, the Observatory's Assistant Director for Science, and the late Carl Tillinghast, the Observatory's young and able Assistant Director for Management, whose untimely death in 1969 was a great loss for the Institution.

Dr. William Melson, Supervisor of Smithsonian's Division of Petrology and Volcanology, was not only instrumental in helping plan the organization and early development of the Center but was responsible for the management of a number of Smithsonian-sponsored volcanological expeditions.

Finally, a host of other individual Smithsonian staff members have offered assistance by evaluating reports, by suggesting imaginative approaches to operational problems, and by providing professional and technical expertise in the fields of communications, data processing, and publications.

Robert Citron
Director, Smithsonian Institution Center
for Short-Lived Phenomena

AN INTRODUCTION TO THE
CENTER FOR SHORT-LIVED PHENOMENA

The space program may have produced a fringe benefit unimagined by the technocrats who promised man the moon: Their flight into space showed us our planet.

On Christmas Eve, 1968, as the Apollo 8 spacecraft sailed over the lunar surface in the first circumnavigation of the moon, the astronauts watched the earth rise over the lunar horizon bright and round as some giant's glassy agate.

From the perspective of 250,000 miles, the good green earth appeared more like a bluish blob of gas, small and lonely in the void of space. Aside from a sobering reminder of our own insignificance, the new view from the moon dramatically showed that the earth, too, is a spacecraft adrift in the universe. Completely dependent on its own self-contained resources, our planet rides as precariously along the edge of disaster as any satellite launched by man.

Ironically, then, at the very moment man gained the moon, his thoughts turned back toward earth. Reaching out for the mysteries of the universe, man discovered the greatest mystery of all: What forces held this spinning clod of dirt and gas together?

For environmentalists, the view of the earth from space became a striking symbol of *the planet as an ecosystem*. For other scientists, the perspective of the lunar view was a vivid reminder of the desperate need for this same type of all-encompassing "overview" of the planet on a continuous basis.

One of the greatest scientific problems of the twentieth century has been finding a means to observe large-scale changes in the earthly ecosystem while they are actually occurring. Many important natural events—volcanoes, earthquakes, sea surges, massive biological changes, and, more recently, oil spills and industrial pollutions—occur suddenly and unexpectedly in remote corners of the world. Often, too, they end as abruptly as they began. Usually the only notifications about short-lived events are the brief reports appearing in the popular press. Yet, while the events may be transitory and short-lived, their effects are sometimes long-lasting. Volcanic islands may erupt, subside, and disappear in a matter of days, but volcanic pollutants may linger in the atmosphere for years, the disruption of local biology may persist for centuries, and the alteration in geological substrates may remain forever.

In 1963, such an undersea volcano erupted off the coast of Iceland with a force unparalleled in recent history. Intense scientific interest was generated by the event, for it marked the birth of an entirely new island—Surtsey.

Surtsey was a scientific bonanza. Iceland was within easy reach of research centers in Europe and America, so hundreds of scientists had a ringside seat for the creation and evolution of a new volcano from beginning to end. Indeed, the study of evolving ecology on Surtsey continues today.

The extensive scientific coverage of Surtsey was possible partly because of its accessibility. However, another major factor was the flow of continuous, accurate, and comprehensive information from Icelandic scientists to their counterparts around the world. It was the availability of information that convinced many researchers to make the trip to Surtsey.

Encouraged by the Surtsey experience, a group of scientists at the Smithsonian Institution, under the direction of Dr. Sidney Galler, then Assistant Secretary for Science, began discussing the possibility of establishing a single world center for the receipt and dissemination of information about unpredictable and short-lived natural events of intense scientific interest. In short, the Smithsonian group wanted a "news service for scientists" that could alert researchers in every discipline to major events occurring at any time in any part of the world.

The Smithsonian Institution was perhaps a natural spawning ground for such a global news service. A host of scientific disciplines—from anthropology to zoology—are embraced by the Smithsonian's broad charter "to increase and diffuse knowledge among men." Although officially a part of the U.S. government, the Smithsonian operates more as a national university than as a federal agency. In fact, the Smithsonian was founded, and is still partially supported, with private money held in trust by the Congress, and thus generally stands apart from the usual political motivations and administration changes.

Also of particular importance to any worldwide organization was the fact that many of the Smithsonian's 13 separate bureaus were already engaged in international operations. For example, the Smithsonian Astrophysical Observatory still maintains more than a dozen field stations around the world for space research, satellite tracking, and atmospheric studies. All these facilities are joined by a global communications network linked to Observatory headquarters in Cambridge, Mass. Moreover, the Observatory operates a worldwide information network for astronomers. The Central Telegram Bureau of the International Astronomical Union, based at the Observatory, serves as the clearinghouse for news of unusual astronomical discoveries and advances.

Dr. Galler hoped that a similar center could be established for natural events affecting the earth, its environment, and its inhabitants. By mid-1967, the Smithsonian Institution decided to set up such a center at Cambridge to take advantage of the Observatory's existing communication lines. Mr. Robert Citron, an Observatory field manager in Africa, was called home to become the first director of the "news service" and to establish a sort of global liaison with the international scientific community.

On January 2, 1968, with one secretary and two phone lines, Citron officially opened what was now called "The Smithsonian Institution Center for Short-Lived Phenomena." He immediately found that a major event already was occurring halfway around the world.

In a tropical replay of Surtsey, a volcanic island had bubbled out of the Pacific in the Metis Shoals area a few hundred miles north of Tonga. One of the Center's first acts, then, was to dispatch (actually, detour) an Australian-bound scientist to the Tonga Archipelago to observe the rare event firsthand.

This short-lived event was *particularly short-lived!* Local reporters claimed the island was already sinking back into the sea. The Smithsonian scientist luckily arrived in time to see the last stages of volcanic activity, to take photographs, and to gather samples of pumice. Information gathered by the one-man expedition was immediately sent to volcanologists and marine biologists around the world.

The almost instant results of the Center's response to the Metis Shoals eruption vividly demonstrated its capabilities and potential. Within six months, the Center had enlisted the aid of some 300 scientists around the world who agreed to serve as "correspondents" for their particular areas. When and if events of scientific interest occurred in their countries, they would notify the Center by telephone, telegram, or cable. In turn, the Center would pass on the reports to any other scientist who had indicated an interest in that particular type of event.

While an occasional Smithsonian scientist may fly off on an expedition to some remote spot, the Center's personnel usually remain at home to contact event areas, obtain information and data on the events, interview reliable witnesses in event areas, request photographic and cinematographic documentation when available, issue event notification and information cards and cables, and distribute the preliminary results of field investigations.

The Center's ability to maintain this long-distance contact with event areas while the events are in progress is made possible through the excellent global communications facility of the Smithsonian Astrophysical Observatory in Cambridge.

This facility operates 24 hours a day and utilizes nine telephone lines and six Teletype circuits, including Western Union Domestic, Western Union International, ITT and RCA International, as well as government networks, thus allowing contact with event areas anywhere in the world, often within minutes.

During the past four years, the Center has gathered and disseminated information on an awe-inspiring list of major phenomena reported by correspondents at the rate of at least one major event every five days. Some of the more "normal" happenings have included bird kills, oil spills, meteorite falls, volcanoes, and earthquakes. Some of the more unusual events have included a mass migration of squirrels in the Eastern United States, a butterfly infestation in Trinidad, an invasion of hungry army ants in Peru, a plague of giant snails in Florida, and the discovery of "stone-age" tribes in South America and the Philippines.

In fact, between 1968 and 1971, the Center reported 71 major volcanic eruptions, 86 major earthquakes, 51 major oil spills, 40 major fireball events, 15 meteorite falls (all had specimens recovered for laboratory analysis), and 137 ecological events including major animal mortalities, migrations, infestations, and population fluctuations, major vegetation mortalities, and major pollution events. The Center also reported other short-lived events such as landslides, storm surges, tsunamis, tidal waves, major floods, forest fires, and several urgent anthropological events (four new tribes discovered).

More important, scientific field teams have investigated over 250 of the 427 short-lived events reported to the Center since 1968.

At the same time, the number of correspondents has grown from the original 300 to more than 2,500 in some 150 countries. Today, the Center has a "scientist-reporter" contact on every continent and on every ocean and in virtually every country of the world. Some of the newer African nations are still unrepresented; and mainland China remains a silent hold-out despite repeated invitations to join the network. By contrast, the Soviet Union has responded enthusiastically to the concept. The U.S.S.R. Academy of Sciences, in a letter to the Center, has noted that "some say this is the greatest scientific information service that has ever been [created]."

This book, however, is really not about the Center, which is only a clearinghouse for information. The Center is not designed to outfit expeditions or conduct research itself (although it occasionally sponsors both); rather, it serves as a catalyst for scientific action. Aside from cutting through bureaucratic red tape and hurdling some occasional nationalistic roadblocks, the Center remains what it was originally designed to be: a news service for the rapid communication of scientific information to researchers studying the world's changing environmental condition.

The Center was established solely to improve opportunities for research through the development of an effective global alert system. Rapid receipt of event information permits research teams, with their instruments and equipment, to enter event areas in as short a time as possible to collect important data that might otherwise be irretrievably lost to science.

This book is most concerned with the events reported by the Center, how they relate both to each other and to similar events in the past and how they contribute to an understanding of the earth. The bulk of this text is devoted to a chronological listing of the events as they appeared in the past annual reports of the Center. In the interests of brevity and clarity, some of the reports have been shortened and simplified. Generally, however, they are reprinted here in almost the same form received from the scientist-correspondents who first reported them to the Center. Readers may interpret this material in any way they wish—as a frightening record of natural disasters, as a chilling indictment of human folly and greed, or as the first "overview of the earth."

At its most basic and appealing level, this book is a fascinating recapitulation of the exciting, interesting, unusual, and often oddball events that occurred in this world during a four-year period.

This, then, is the Smithsonian Institution's "state-of-the-planet report" to the global community.

EVENT REPORTS   1968

Metis Shoal volcanic eruption, Tonga Islands, southwest Pacific. Aerial view.

1. DECEPTION ISLAND VOLCANIC ERUPTION
South Shetland Islands **December 1967**

Violent tremors and three explosions in Telefon Bay, Deception Island, Antarctica, provided the first short-lived event ever recorded by the Center for Short-Lived Phenomena. It was a happy coincidence that the first report concerned a birth.

Deception Island volcanic eruption, South Shetland Islands, Antarctica. A scientist collects samples of blocks and ash on the new island.

Beginning in December 1967 and continuing through January 1968, a new island was born in Telefon Bay. The island measured 900 meters long, 250 meters wide, and 120 meters high.

Metis Shoal volcanic eruption, Tonga Islands, southwest Pacific. Aerial view.

2. METIS SHOAL VOLCANIC ERUPTION
Tonga Islands **December 1967**

Five days after the birth of an island in Antarctica, another island was born in the southwest Pacific, but on a more spectacular scale. The island that thrust out of the Pacific with almost mythological fury remained above water for only two months, barely enough time for scientists to collect samples.

The short, violent life of the Metis Shoal provides a fascinating biography. A submarine volcanic eruption at Metis Shoal pushed itself above the surface on December 12. The next day, a report described the eruption as taking the form of "an incandescent island" about one-half mile long and 150 feet high glowing "cherry red at constant intensity." A pillar of steam and smoke surged from the sea to heights of 3,000 feet. And, out of the chaos of the island's birth, boulders of molten lava were ejected to heights of 1,000 feet.

Eruption activity lasted approximately 27 days; and the island itself stayed above the surface for 58 days.

The island, formed after 20 days of submarine activity, was about 700 to 800 meters long, 100 to 150 meters wide, and some 15 to 20 meters high. By late December, the island was a circular mass of gray rocks with streaks of red flame glowing through the steam and smoke that poured into the air. From 17 miles away, flames were seen leaping above the horizon. Ten miles distant, the volcano was a phantasmagoria of flames and lava fountains shooting over 1,000 feet into the night sky. From its single cone, huge billowing clouds of white smoke were clearly visible at night. Hundreds of tons of rocks were thrown into the air, and the entire area smelled of sulphur. The water around the island turned a strange emerald green streaked with brown.

Yet, once the volcanic activity ceased in early January, the island began to erode quickly. By January 25, it had eroded to what one observer called "rocky outcrops."

By the second or third week of February, it had disappeared entirely, leaving only two clues to show that it had ever existed: very high waves breaking over submerged rocks, and then, after those had subsided at the beginning of April, only a brown discoloration of the water where the island had once been.

While en route to Australia, Dr. Charles Lundquist of the Smithsonian Astrophysical Observatory visited the Tonga Archipelago to photograph and collect samples of the Metis Shoal. He collected fresh rock specimens from the summit of the volcano in mid-February, well after the volcanic activity had subsided. The summit was then four to five feet beneath the surface of the sea. Because of the Center's quick response to this event—even though the

island stayed above the surface for only eight weeks—it was possible to document the event and make detailed analyses of the volcanic material.

Metis Shoal volcanic eruption, Tonga Islands, southwest Pacific. Aerial view.

3. SICILY EARTHQUAKE
Sicily, Italy   **January 15**

A total of 252 people were killed, approximately 1,500 were injured, and 83,000 were displaced from their homes in one of the most tragic natural disasters in recent years. The deadly quake occurred suddenly on January 15, but a number of aftershocks were recorded for several days following, and tsunami waves were reported on the west coast of Sicily near Trapani.

4. POLO FIREBALL
Polo, Illinois   **February 26**

The first of many fireballs reported by the Smithsonian Center was seen traveling across the north-central United States sky by hundreds of people, including airline pilots. The object, visible in midafternoon, was described as bright green and appeared to explode into three separate objects. Some observers claimed to have seen it strike the earth, and requests for additional information were broadcast on local radio stations and published in local newspapers at the suggestion of the Center. Although reports continued coming in, it became clear by the afternoon of February 29 that the object had not reached the earth intact; or if it had, fragments would have been too difficult to locate.

5. *OCEAN EAGLE* OIL SPILL
San Juan, Puerto Rico   **March 3**

San Juan harbor and several miles of nearby beaches were fouled by some 5.7 million gallons of crude oil that oozed from the wreckage of the tanker *Ocean Eagle,* literally split in half by pounding seas at the entrance to the harbor. Populations of waders and plovers were thinned out, over 500 pelicans were pulled out of the oil by volunteers, and the harmful effects on marine life in general in and around the San Juan harbor were considerable.

The day after the spill, the U.S. Navy and Coast Guard as well as the University of Puerto Rico's Department of Marine Sciences began an investigation of the effects of the oil—and the detergents used to emulsify the oil—on marine life and ecosystems.

6. DAYTON FIREBALL
Dayton, Ohio   **March 3**

Although called the Dayton Fireball, this meteor was reported by observers over a wide area including Columbus, Cincinnati, Cleveland, Louisville, Ky., and even as far away as Atlanta, Ga. No meteoritic material was recovered as a result of the sightings.

7. VERACRUZ FIREBALL
Veracruz, Mexico   **March 27**

Three huge meteorites turned the night into day over Veracruz, Mexico, and shook this port city when the brilliant objects reportedly crashed into nearby hills. Some observers noted "an intense light changing from blue to red to a dazzling white while crossing the sky." Indeed, the Veracruz Fireball is probably one of the most spectacular—and colorful—fireballs recorded. One report called it "a brilliant globe with a trail of sparks behind accompanied by a buzzing sound." Fishermen on the Gulf of Mexico said the fireball "began as greenish stars which turned the night sky brilliant white, and then became blue before impact."

8. SCHENECTADY METEORITE
Schenectady, New York   **April 12**

Appearing to observers as "half the size of the full moon," a fireball streaked across the skies of central New York State early on the morning of April 12. At least one fragment of this fireball reached earth—crashing into the roof of a house in Scotia, N.Y.

The meteorite, which struck the house with a loud bang "like a firecracker," was a small stone weighing 283.3 grams. Another small

Ocean Eagle oil spill, San Juan, Puerto Rico.

piece apparently broke off on impact and was not found.

The Smithsonian Center, although not learning of the meteorite's recovery until a month after its fall, later conducted a full-scale investigation. Samples of the meteorite were obtained for radioisotope analysis at the Smithsonian Astrophysical Observatory in Cambridge, Mass. The main body of the meteorite became part of the specimen collection of the Schenectady Museum.

Schenectady meteorite, Schenectady, New York.

9. MOUNT MAYON VOLCANIC ERUPTION
Philippine Islands   **April 20**

Mount Mayon, an 8,000-foot-high conic volcano on the island of Luzon just 200 miles southeast of Manila, had been quiet since 1947. But on April 20, the awesome volcano erupted in a terrifying display of power, spewing steam and molten lava from its crater mouth and shooting balls of fire 3,000 feet into the air. By the next day, the volcanic activity had become so intense that the 70,000 residents living within a six-mile area of the volcano were evacuated.

The Mt. Mayon eruptions actually spanned a period of many days—beginning with initial explosions on April 20, building to a peak volcanic activity in early May, and continuing with weak sporadic ash explosions and lava extrusions until the end of that month.

Lava flow began in the early morning of April 23, and by the next day, a mammoth,

30,000-foot pyroclastic cloud (containing solid volcanic particles) had risen from the cone. By then, 20 separate explosions had been recorded. These explosions continued beyond May 2 when individual explosive events reached a peak.

When the volcano erupted, strong lava fountaining was observed accompanied by a quiet overflowing of lava from the crater between eruptions. Carefully observed—and photographed—was the phenomenon called *nuées ardentes*, literally, *burning clouds*.

Mount Mayon turned out to be one of the most fully documented volcanoes in history. The Center, which issued 35 reports on the life of the volcano from April 21 to November 8, maintained daily communications with scientists and journalists in the area. Data on frequency of eruptions, eruption time, magnitudes of eruptions, lava flows, ash falls, and pyroclastic cloud descriptions had been recorded and relayed by the Center to its subscribers. On April 24, the U.S. Air Force began systematically taking aerial photographs of the volcano—an assignment that extended over a period of the next six weeks.

A Center-sponsored team spent nine days on the volcano observing eruptions, collecting samples, and directing a film crew. That film, as well as the more than 2,000 high-resolution aerial photographs received by the Center, provides valuable insights into the awesome mechanics of volcanoes.

10. TOKACHI-OKI EARTHQUAKE
Hachinohe, Japan   May 16

The Tokachi-Oki earthquake, with a Richter scale magnitude of 8.25, was the strongest to hit Japan since September 1, 1923, and the strongest worldwide since the great Alaskan earthquake of 1964. With its epicenter located under the sea floor 100 miles east of the city of Hachinohe, the quake caused tsunamis—which means *big waves* in Japanese—measuring as high as 2.8 meters along the north coast of Honshu, the major island in the Japanese chain.

The tsunamis were responsible for the sinking of at least 33 vessels and for the flooding of hundreds of homes. Two 7,000-ton ships were driven into coral reefs; and, in the Hokkaido-Tohoku regions, a total of 62 ships, including 51 fishing boats, were lost, sunk, or capsized by high waves. Eight other freighters

and fishing vessels ran aground or were capsized in the quake's aftermath.

The earthquake left huge gaps in the earth, cut communication lines, and halted transportation on Hokkaido and northern Honshu islands, while aftershocks cracked or damaged roads at 228 different locations.

11. INANGAHUA EARTHQUAKE
New Zealand   May 23

Exactly one week after the Tokachi-Oki earthquake rocked Japan, another quake struck another large Pacific island chain—New Zealand. The Inangahua earthquake, with a magnitude of 7.0, struck western New Zealand, shifting bridges, cracking roads, causing landslides, and damming rivers.

Several dozen very strong aftershocks—each greater than magnitude 5—occurred from May 24 to 26, causing additional and widespread destruction over acres of New Zealand countryside.

12. FERNANDINA CALDERA COLLAPSE
Galapagos Islands   June 11

The Galapagos Islands, a chain of small islands about 650 miles west of Ecuador, are famous to scientists and laymen alike because of their unique animal and plant life. Charles Darwin, who visited the islands in 1835, found a living laboratory to test some of his principles of evolution. In June 1968, 133 years after Darwin's visit, one of these strange, primeval islands was threatened by volcanic activity, and if it were not for a freakish volcanic reversal, some of the animals and plants Darwin used as theorem proofs could have been destroyed.

The volcanic activity in the uninhabited island of Fernandina was internalized, when the caldera, or crater floor, collapsed, lowering approximately 1,000 feet. The collapse took place on July 4, a little less than a month after eruptions began. The initial shock in June was recorded by infrasonic detection stations in North America and described as "stupendous" and "in the multi-megaton range." Later tremors were recorded in the 4.5 to 5.5 Richter magnitude range.

The volcano offered a "multi-megaton" visual spectacle, as well, to the people of Santa Cruz Island some 80 miles away. A "mushroom-shaped cloud" rose in an otherwise clear sky.

Mount Mayon volcanic eruption, Philippine Islands. Mount Mayon volcanic eruption showing *nuée ardente* phenomena on April 27, 1968. Pyroclastic eruption cloud reached height of 30,000 feet.

(That cloud eventually drifted over Santa Cruz itself.) Electrical flashes over the volcanic area lit up the night sky and a fallout of gray volcanic ash was carried 50 miles from the volcanic source.

Seismic activity continued for several days after the first day's fireworks, but the volcanic activity subsided enough to allow a party of scientists from the Santa Cruz Darwin Station to venture to the rim of the Fernandina caldera. But, as the men approached the rim, violent tremors shook the entire island. During a six-hour visit, 56 tremors were counted, each lasting from two to six seconds, with 14 powerful enough to cause rock falls on the nearby slopes. Because the tremors came in waves, growing and subsiding, instead of in abrupt staccato shocks, the Santa Cruz party got the

The scientists most feared for the loss of flora and fauna. The crater had been no barren and lifeless pit atop a mountain, but rather a sheltered haven for life on the island. In its center there had been a lake, complete with islands and inlets due to the unevenness of the caldera floor. The lake was bordered with vegetation, mostly reeds that supported the largest population of the Galapagos duck in the archipelago (some 1,900 according to a February 1968 count) as well as other aquatic birds, including the black-necked stilt. There was also a small forest near the crater floor and on the less steep slopes of the inner walls.

Because of the clouds of dust, the Santa Cruz scientists could not easily see the floor of the crater, but they judged that the most active area was the southeastern part of the caldera

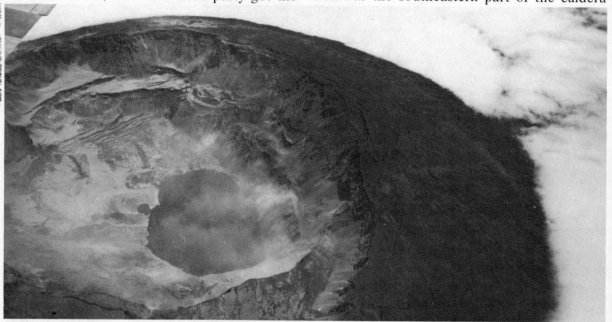

Fernandina caldera collapse, Galapagos Islands. Aerial view.

weird impression that the whole island was balanced on a "jelly-like" substance.

These undulating movements affected the inner walls of the caldera, causing tremendous rockfalls at frequent intervals and giving rise to massive clouds of dust that filled the giant crater and flowed over the rim to be carried away by the wind. The scientists likened the sound of the falling rocks to the roar of turbulent seas breaking on a rocky beach. After each landslide, a powerful gust of air soared up from the crater floor.

Through the dust, the scientists could see that large areas of the crater floor had disappeared. Those areas still remaining intact were discovered to be heavily fissured, cracked by the intensity of the volcano's initial blasts.

where there were fumaroles, or holes through which volcanic gases are emitted.

In July, aerial photographs showed a drastic shrinkage in the size of the lake. Before the volcanic activity, the lake covered the entire floor; but, in July, with the caldera floor collapsed, the rim walls and plateaus crumbled, the lake was more like a pond, tucked into a corner of the crater. But little damage outside of the crater itself could be seen.

Although the entire island was covered by a non-toxic volcanic ash, the unique plant and animal life of the island apparently had been spared by Fernandina's unique type of volcanic activity. Instead of spewing red-hot lava, crushing boulders, smothering ash, and poisonous gases from the bowels of the earth, the

Fernandina caldera collapse, Galapagos Islands. View from east rim.

volcano collapsed in upon itself. This "reverse volcano" was the result of a drastic fall in internal pressure rather than the usual build-up. The result was the containment of all destruction within the crater and the salvation of the rest of the islands.

Nevertheless, the ecology of the crater was disturbed by the eruptions. Much of the vegetation that had covered the walls of the crater disappeared. Many of the crater's young and nesting birds were apparently killed. A once secret and peaceful hideaway teeming with life had vanished.

13. *WORLD GLORY* OIL SPILL
Durban, South Africa   **June 14**

A raging storm split the oil tanker *World Glory* in half 90 miles off the coast of Durban, South Africa. Some 50,000 tons of crude oil gushed from the belly of the crippled ship. The oil pouring from the ruptured tanker was pushed shoreward by high winds and seriously threatened the marine and wildlife of the St. Lucia Game Reserve.

On June 17, three days after the tanker break-up, the oil slick was reported within two miles of the coast. Authorities feared the black mass might reach the St. Lucia estuary and contaminate the shoreline, harming marine bottom life, marine bird populations, sardine schools, and other aquatic life there.

To counter the oil scourge, a massive operation was begun to spray the area with detergents. Fly-ash residue from burnt coal was also used to soak up the oil and precipitate it to the ocean bottom. By June 24, it was reported that the spraying operation, enormous in scope, had effectively dispersed the spilled oil.

By the end of the month, South African officials could happily report that the spraying operation was more effective than they had first realized and that none of the spilled crude oil had reached the shore.

14. MOUNT ETNA ERUPTION
Sicily, Italy   **June 16**

Probably the most well known volcano in the western world, Mt. Etna, made headlines in June when the international news services reported that earth tremors had been felt in the vicinity of the mountain and that streams of red-hot lava were pouring down its slopes.

Mount Etna actually had been active since January 1967. This activity could be observed when the vent of the northeast crater of the volcano was opened to expose a column of fluid magma. Gases escaped through the crater, producing intermittent explosions that spewed lumps of liquid lava into the air—sometimes as high as 600 feet.

The intervals between explosions varied from a few seconds to several hours and sometimes to several days. For a while, only smoke issued quietly from the crater, until a new, somewhat stronger explosion spit out ash clouds, presenting and impressive spectacle, particularly at night.

From an investigation made in June 1969, it was reported that new lava channels were being formed, while older channels were not being renewed. The scientists said in time a great lava cake will have formed, covered by a crust. Through this crust, new secondary boccas or vents will form, from which small lava flows can escape. Such an activity, the scientists said, may go on for years without any kind of true eruption. The only evidence of actual activity on Mt. Etna is the existence of a large blowhole in the western part of the central crater from which roaring gases escape violently at short intervals.

15. MOYOBAMBA EARTHQUAKE
Moyobamba, Peru   **June 19**

Houses crumbled into the shaking earth, trees were uprooted, 11 people were killed, and hundreds more were injured in a powerful and disastrous earthquake centered near Moyobamba City, and aftershocks of great strength were reported for several days following the initial quake.

16. FLORIDA FISH KILL
Florida   **June**

Although Hurricane Brenda did not strike the United States mainland in late June, it indirectly caused the deaths of thousands of tropical fish along a 70-mile stretch of Florida coastline. The powerful spiral of the hurricane sucked an icy underwater tide from the depths of the Atlantic and redirected it toward the Florida coastline. This chilling undercurrent proved deadly for many species of fish living in the warm waters of the Gulf Stream reefs.

Stunned by the cold Atlantic current, brightly colored fish died by the thousands and their bloated bodies covered the surface of the sea. Hundreds of pounds of angelfish, trumpet, file fish, cardinal fish, triple tails, and reef bass were picked up at sea. None of the carcasses were washed onto Florida beaches due to strong seaward winds.

The Florida Board of Conservation reported that the temperature in the reefs dropped from 65 degrees at the water's surface to 43 degrees 30 feet down.

17. CALIFORNIA FISH KILL
Stanislaus River, California   **June**

At almost the same time that cold water in the Gulf Stream was killing hundreds of fish off the Florida coast, another fish kill was occurring on the other side of the country. However, the reason for the deaths of an estimated 100,000 fish in the Stanislaus River was much more terrifying than any freak of nature.

The vast number of sturgeon, carp, catfish, striped bass, smallmouth black bass, shad, bluegills, hardheads, and sunfish found dead

in the river posed a mystery for California officials. The lack of any identifiable natural causes for these mass deaths, however, led the California State Fish and Game Department to consider the inevitable—and unnatural—cause: the pollution of the river water by pesticides, herbicides, and other lethal man-made substances.

18. VERMONT WINDSTORM
Underhill, Vermont   **June 27**

On June 27, strong southeast winds with gusts up to 55 miles an hour caused extensive damage to hardwood trees in the Underhill area. Most seriously hit were those trees on slopes exposed to the unrelenting gusts of the storm.

Dr. Frederick M. Laing of the University of Vermont investigated the effects of the windstorm on the foliage. He also began long-range studies of "dieback," a plant disease in which branches or shoots die from the tips inward, and other diseases that might be traced directly to the storm.

19. HUNTINGTON FIREBALL
Harrisburg, Pennsylvania   **July 2**

A *bolide* is a fireball with a difference: it explodes. The bolide known as the Huntington Fireball exploded into a number of particles, and the sound of those explosions, plus accompanying sonic booms, was heard by thousands of people from Akron, Ohio, to Harrisburg, Pa.

The fireball traveled in a northerly direction across west-central Pennsylvania and appeared as an incandescent ball. According to one observer, "it lit up the whole countryside." The last sighting of the meteor was reported from Binghamton, N.Y. No meteoric material was discovered.

20. NEWCASTLE MUSSEL POISONING
Newcastle, England   **June – July**

Beginning in June and extending into July, an outbreak of mussel poisoning was reported in Newcastle, England. The outbreak was closely watched by a number of research groups, which determined that the poisoning might have been caused by a dinoflagellate bloom on the northeast coast of England during the first half of May.

Dinoflagellates, plantlike protozoa with two flagella, or whiplike appendages used for locomotion, are important elements of plankton. The dinoflagellate allegedly causing the outbreak was never positively identified.

21. VALPARAISO SEA SURGE
Valparaiso, Chile   July 25–26

Suddenly and without warning, tidal waves 15 to 18 feet tall rolled into the Chilean coastline near the capital city of Valparaiso on July 25 and 26. The huge waves were apparently caused by a sea surge far offshore. Although much coastal property and many ships were damaged, there were no personal injuries reported.

The question that needed answering, however, was what caused the sea surge? Some reports indicated the waves were associated with seismic phenomena; other theories attributed the waves to a storm far out at sea. Today, however, their origin remains a mystery.

22. OMAHA FIREBALL
Western Iowa and Nebraska   July 28

With the sun still high in the sky during the day on July 28, people living near Iowa City, Leon, and Des Moines, Iowa, and Omaha, Neb., reported observing a brilliant fireball. Many observers also reported seeing a contrail and smoke trail several minutes after the meteor had passed. Although information on the fireball and the possibility of a meteorite impact with earth was disseminated through radio broadcasts and newspaper publications around the event area, no meteoritic material was found.

23. MOUNT ARENAL VOLCANO
Costa Rica   July 29

Successive violent and devastating eruptions of the Arenal volcano in northwestern Costa Rica wreaked havoc, death, and destruction over many acres on the western slopes and flank of the mountain. Arenal's *nuées ardentes*, or glowing-cloud type, eruption produced deadly blasts of hot gases, ash, and incandescent blocks of rock and lava. The rocks and lava rolled swiftly down the slopes of the volcano, crushing everything in their path. The rocks left huge impact craters, similar to those created by bombs.

The explosive eruptions subsided by August; however, the countryside was already covered with heavy ash from the explosion clouds. In addition, the volcano's fumarolic activity continued well into September. On September 20, new lava flows from the crater began. Finally, in November, a lava stream about 600 feet wide and about 40 to 60 feet high was observed moving down the mountain at a speed of approximately six feet per day.

Lava continued to flow from Arenal throughout the year and continued into 1969. The activity was usually accompanied by vapor eruptions, but on May 3, 1969, simultaneous activity was reported at three Costa Rican volcanoes—Arenal, Poás, and Rincón de la Vieja.

On May 16 aerial investigation revealed considerable lava activity and avalanches on the upper western flank of the main lava flow. Activity continued through the summer and, by mid-September, the lava flow had advanced a quarter of a mile farther, severing a road between La Fortuna and Pueblo Nuevo.

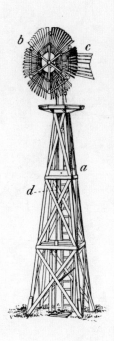

Mount Arenal volcano, Costa Rica. Destruction of land caused by volcano.

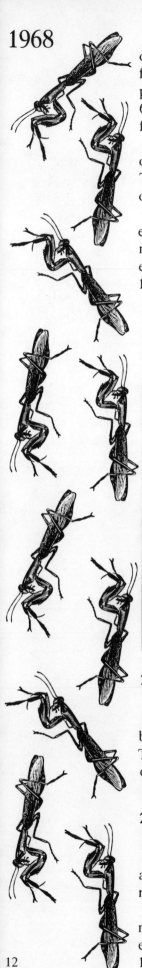

Also in September, another lava flow moved on a broad front west and southwest of the first flow at a rate of approximately five feet per day. The head of the fresh lava flow was 60 to 100 feet high no more than two miles from the lower crater.

At the end of 1969, there was a resurgence of the activity and the rate of flow increased. The flow had built a continuous mass of lava one mile long and a quarter-mile wide.

Ironically, the 1969 flow completely surrounded a hill on which a monument stands as a memorial to the 80 persons killed during the explosive phase of Arenal's eruption in July 1968.

26. SUDAN LOCUST SWARM
Republic of Sudan   **August 6**

In August, the African republic of Sudan experienced its worst locust swarm in ten years. The insect plague infested, according to the Food and Agricultural Organization of the United Nations, over 250,000 square miles of land and threatened the Sudanese cotton crop.

When a mass effort of spreading insecticides from aircraft was finally started, the locusts had already settled into the area and had begun breeding.

The intensive air and ground control operations greatly reduced the insect population in the infested areas, but it wasn't until late

Mount Arenal volcano, Costa Rica. Impact crater and partially destroyed house.

24. MANILA EARTHQUAKE
Philippine Islands   **August 1**

A massive earthquake toppled multi-storied buildings in downtown Manila on August 1. The quake caused millions of dollars' worth of damage and killed several hundred people.

25. OAXACA EARTHQUAKE
Oaxaca, Mexico   **August 2**

One day after an earthquake rocked Manila, another shook the southwest coast of Mexico, near the state of Oaxaca.

The Mexican quake was strong, registering magnitude 7.5 on the Richter scale, with an epicenter located in the Pacific Ocean about 175 miles southeast of Acapulco.

October that the interior of the country was clear of most swarms.

27. VIRGINIA FISH KILL
Richmond, Virginia   **August 12**

Because of the introduction of toxic man-made materials into the James River below the city of Richmond's deep-water terminal, a number of fish were killed. Although the total number of fish deaths was unknown, the Virginia State Water Control Board said the number was significant.

28. NORTHERN CELEBES EARTHQUAKE
Indonesia   **August 14**

The first reports from the Northern Celebes Sea claimed that a very strong earthquake, with a 7.4 Richter scale magnitude, had caused a small island off the coast of Indonesia to crumble into the sea.

The reports were erroneous. The island had not vanished, but the earthquake had taken a terrible toll. The quake caused considerable damage on the island, killing an estimated 200 persons and damaging several hundred homes.

29. COMODORO RIVADAVIA FIREBALL
Southern Argentina   August 17

Personnel at a Smithsonian Astrophysical Observatory satellite tracking station in Comodoro Rivadavia, Argentina, reported observations of a large, bright, flying object accompanied by at least 14 other smaller ones. The objects were observed for about half a minute, during which time they appeared to change color from orange to blue-green and left a flaming trail of sparks and smoke.

The SAO observers at first thought they had seen a meteor, and a spectacular one at that. But the "meteor" was later identified as the re-entry and death of a Soviet satellite, the Cosmos 211, launched on April 9, 1968.

30. CHILEAN DROUGHT
Chile   March–August

The people of Chile have been waging war with their environment since the republic first came into being. In 1960, earthquakes, tidal waves, and volcanic eruptions caused the deaths of 5,000 people. Eight years later, another kind of natural disaster—less tumultuous, but no less pernicious—struck the country.

The Chilean drought extended for eight months and was regarded as the worst in 120 years. More than half of Chile's nine million people were directly affected. Some 150,000 sheep died of thirst. About 116,000 square miles of once lush agricultural land was made barren by the scarcity of water. And three provinces were declared disaster areas.

The long-term effects of the drought are still felt by the country's economy and biological environment.

31. KILAUEA VOLCANIC ERUPTION
Hawaii   August 21

Hawaii, like many other Pacific islands, was formed when a huge underwater volcano spewed lava to the surface where it solidified as the tenuous terrestrial tip of a deep-sea mountain. Some of these ancient marine volcanoes are still very much alive.

On the grounds of the Hawaii National Park, on the island of Hawaii, two volcanoes, Kilauea and Mauna Loa, have active craters that have erupted many times during this century. On August 21, the Kilauea volcano erupted with fountains of lava up to 75 feet.

Perhaps a more interesting event took place four miles south of the park, where activity broke out in the Hiianka crater, which had not been active since prehistoric times.

At 2:35 p.m., October 7, the Kilauea volcano erupted again after three and a half hours of harmonic tremors and many earthquakes. The eruption created fissures two to three miles long that crossed Napau crater on the east rift zone, nine miles southeast of the summit.

32. ALBANY FIREBALL
New York State   August 26

The Albany Fireball was observed by thousands of people in the northeastern United States on August 26. Reports from Maine, Connecticut, Massachusetts, and New York described the object as having a "thick yellowish tail" and to be "as bright as a drop flare." Traveling in a westerly direction, the fireball provided splendid entertainment for those who viewed it. One observer said it displayed three distinct colors—orange, yellow, and red—as it flew across the sky; when it terminated, the meteor went out in a blaze of glory, with flames of red and green.

The last person to spot the meteor was a pilot flying over Albany, N.Y. From eyewitness reports and some limited photographic data, the meteor's trajectory was roughly sketched, with an end-point presumed to be somewhere near the northern end of Lake Champlain close to the Canadian border.

No ground search for the meteorite was undertaken because there was a lack of definite information on a probable impact point.

33. GRAND RAPIDS FIREBALL
Central Michigan   August 27

A fireball that streaked across central Michigan at midday August 27 was bright enough to be seen in the daylight. Moving north to south, it broke into three or four pieces, leaving a diffuse trail that dissipated very quickly. The

meteor created sonic booms that were heard over the area. One observer reported the sound as "something rushing through the air." No meteoritic material was found.

34. ACCABONAC HARBOR FISH KILL
Long Island, New York　　　**August 29**

About a thousand small fish of various species were found dead in the Accabonac Harbor at Riverhead, Long Island, on August 29. A field investigation of the incident was made by the New York State Bureau of Marine Fisheries. The cause of the fish kill was attributed to the spraying of pesticides in the area.

35. NORTHEAST IRAN EARTHQUAKE
Northeast Iran　　　**August 31**

On August 31, a strong earthquake caused a 40-mile-long rupture in the earth's crust in northeastern Iran. One of the worst quakes in the modern history of the nation, the disaster took the lives of 6,000 people and destroyed 60,000 homes. The quake was recorded as magnitude 7.8 on the Richter scale.

Besides the main rupture, there were fault branches northwest of the affected area with maximum horizontal displacements of 15 feet. Vertical movements showed scissoring in the west part of the fault. In the valley south of the rupture extensive ground fractures of neotectonic origin were detected.

36. BANUA WUHU VOLCANIC ERUPTION
Indonesia　　　**September 6**

The eruption of a submarine volcano, Banua Wuhu, was reported on September 6 by the Indonesian Volcanological Service. The volcano is located near the island Mahengetang, which is part of the Sangihe Islands in the Celebes Sea off the coast of Indonesia. Initially, it was not possible to determine the exact extent of the volcano or determine whether the submarine volcano had built itself above the surface.

Cable reports in the next few days from the American Embassy at Djakarta indicated the major volcanic activity had subsided quickly, and continued in a solfataric stage, with only sulfurous gases, steam, and other gases being emitted.

37. APPALACHIAN SQUIRREL MIGRATION
Southeastern U.S.A.　　　**September**

The gray squirrel is the garden variety of squirrel; the kind Americans feed nuts to in parks or watch frolicking in their backyards. In September, these animals began a massive migration that, as one observer put it, "started so slowly it was hardly noticeable."

The migration quickly picked up momentum, however. By mid-month, motorists were reporting that unusual numbers of dead animals were littering highways in the southern Atlantic states. In fact, some experts estimated approximately 1,000 times as many squirrels as normal were being killed on East Coast roads.

The activity became evident when thousands of gray squirrels were reported scurrying lemming-fashion across open land, backyards, and even lakes and rivers throughout the Appalachian region of Maryland, Tennessee, Virginia, and the Carolinas. The animals were reportedly sick, starving, and deranged—and supposedly headed north to greener pastures. (Newspaper reports somewhat exaggerated the actual situation, however, so that animal lovers throughout the southeastern United States launched various campaigns to feed the squirrels. On the other hand, several sporting groups urged state game wardens to open squirrel hunting season a few weeks early.)

In truth, however, the squirrels were neither hungry nor sick, nor even really "migrating."

The word "migration" was misleading, for the squirrels had no real destination in mind. Not being social animals, they were acting individually. Thus, the migration was more properly described as an "irruption," a mass panic in which squirrels went willy-nilly in all directions.

The cause of the irruption, however, remained a mystery. First reports cited the insufficient production of acorns or the unusually large squirrel population as probable causes —or even a combination of both factors. The hypotheses seemed to make sense until it was later reported that, although the nut and acorn crop was not as heavy as usual, there was no shortage of food. Moreover, the dead squirrels found in the forests and on the highways showed no signs of malnutrition.

The report from a University of Maryland scientist finally provided the best explanation for the squirrel behavior. A bumper mast crop in 1967 had produced two highly successful breeding seasons, in the summer of that year and in the following spring. "The great influx of squirrels, for some reason, triggered an unusual movement of squirrels from area to area,"

Appalachian squirrel migration, southeastern U.S.A.

he wrote. "This extraordinary activity seems to be a widespread phenomenon. Because it occurs at a time of year when food is abundant, there is no relationship to food conditions."

Squirrel irruptions have been noted before, of course, especially during the early part of this century before the great eastern forests were cut down. In fact, such irruptions were quite commonplace and usually occurred in September. Because of the gray squirrel population boom, the 1968 irruption was large enough to be noticed by the public. This does not mean the panic and frantic behavior were caused by overcrowding.

One theory relates the apparent panic behavior of the irruption to the gray squirrel's method of food storage. This species, unlike other squirrels, does not store food in a specific place such as a tree hollow. Instead, this squirrel simply buries mast and other bits of food under a few inches of dirt in a number of places.

The search for new hiding places and the disorientation caused by moving into alien territories might simply have created confusion.

The squirrels' irruptive activity reached a peak during the last two weeks of September. Sporadic reports of squirrel movements were received throughout the month of October; but, by the beginning of winter, the squirrels appeared to have returned to normal.

38. SUBMARINE LAND AVALANCHE
New Guinea   September 17

Although avalanches are usually connected with mountains, they can occur under the sea as well.

On September 17, a submarine earthquake was reported in the Solomon Sea, 250 miles southeast of Madang, New Guinea. The earthquake caused a sand avalanche underwater that might have gone unnoticed had the massive movement of sand not sliced a submarine telephone cable.

The cable, connecting Singapore and Australia, was promptly repaired. Unfortunately, no investigations of the submarine avalanche itself were conducted.

39. MISSISSIPPI FLYWAY
DUCK MIGRATION
Mississippi River Valley   September

The Federal Aviation Administration was understandably concerned. According to ornithologists, the Mississippi Flyway, that is, the route ducks use for their migration from Canada south, would be jammed with the birds in greater numbers than usual in autumn 1968.

The FAA thus warned pilots that 10 million ducks plus 500,000 Canada geese and 450,000 blue geese would be journeying down the Flyway during September and October. Because of the warnings, pilots were careful to give any duck the right of way; and transportation, both duck and human, continued without mishap.

40. VENEZUELA EARTHQUAKE
Caracas, Venezuela   September 20

A strong earthquake, magnitude 7 on the Richter scale, struck the eastern part of Venezuela, causing a considerable amount of property damage. The Center kept in contact with scientists in Venezuela during the critical days

15

of the quake to assess damage to homes and environment.

41. GUARATUBA SUBSTRATA COLLAPSE
Guaratuba, Brazil   **September 23**

Just three days after an earthquake struck eastern Venezuela, part of the resort town of Guaratuba in Brazil collapsed and fell into the sea. Destroyed in the sudden catastrophe were 20 buildings, including the town hall.

According to field investigations quickly undertaken by the government of Brazil, the collapse was apparently caused by sea water seeping into the land under the resort city and weakening the substrata more than 500 feet from the shoreline.

42. RAOUL ISLAND EARTHQUAKE
Kermadec Islands, Pacific   **September 26**

A fairly strong earthquake, recording 6.9 magnitude on the Richter scale, struck the Kermadec Islands in the Pacific. The quake was recorded by seismic stations in New Zealand and Australia and was reported to the U.S. Coast and Geodetic Survey. No investigation of the quake was undertaken.

43. TAAL VOLCANO ACTIVITY
Philippine Islands   **October 7**

A few months after the Mt. Mayon volcano erupted in Luzon, the Taal volcano entered into an active period on October 7, with seven volcanic earthquakes and tremors reported within a 24-hour period. The Taal volcano, one of 12 in the Philippines, is under constant surveillance by the Philippine Volcanology Commission.

44. AUSTRALIAN EARTHQUAKE
Western Australia   **October 14**

The Center kept in contact with Australian information sources from the day the earthquake struck, October 14, until the last tremors were felt on October 31. The quake, Richter magnitude 7, was felt over some 1,600 miles of sparsely populated area in western Australia. The most severely affected area was the town of Meckering, some 20 miles east of Perth, where a number of buildings were badly damaged and many people were injured by falling debris.

45. CALIFORNIA SEAL COLONIZATION
San Miguel Island   **October 14**

On the island of San Miguel, off the California coast near Santa Barbara, two scientists discovered a new colony of immigrants the census takers overlooked. The immigrants were fur seals who had deserted the cold lonely waters of the Bering Sea during the summer of 1968 and moved into the warmer neighborhood of San Miguel.

Most important, the new seal colony was not made up of transients. These seals had apparently come to stay. Already they were breeding new generations of California-based seals. As many as 150 seals were thought to have made the move from Alaska to California. At San Miguel the colony consisted of 60 cows, 36 pups, and one very tired adult male.

46. NEW GUINEA EARTHQUAKES
New Guinea   **September–October**

Between September and October, the remote, largely primitive island of New Guinea suffered not one, but two major earthquakes. The first, with a Richter magnitude of 6.6, occurred on September 8, with the main shock estimated at a depth of about 20 miles. The second quake occurred a month later and was slightly weaker, registering 6.4 Richter. Both quakes occurred in the East Sepik region of New Guinea and caused extensive injuries and widespread property damage.

47. CHINANDEGA FOOTPRINTS
Nicaragua   **October**

The road between the Nicaraguan city of La Mora and the slopes of the San Cristobal volcano near the city of Chinandega was more like a ditch than a highway. There was a six-foot difference between the original roadway and its present level. The road had grown deeper and deeper over the years through the combined action of tropical rains and oxcart traffic. In October, the rain again eroded a little bit more of the road and uncovered a major anthropological find—the footprints of prehistoric man.

The prints, not much smaller than those left by modern men, had apparently been made in prehistoric times and covered by an ash flow from the nearby volcano. Because the footprints were subject to further erosion, investigations commenced almost immediately.

48. POCAHONTAS FIREBALL
Pochahontas, Iowa   **October 21**

The Smithsonian Astrophysical Observatory

operates a network of special automatic cameras in the Midwestern United States for the purpose of photographing bright meteors and recovering freshly fallen meteorites.

On the night of October 21, the Smithsonian's "Prairie Network" photographed the Pocahontas Fireball, a very bright meteor seen over Iowa. Using the eyewitness reports of the fireball, reports of associated sonic booms, and the photographic data, Smithsonian scientists computed a probable impact point near Pocahontas, Iowa. Several specimens were sent to the Smithsonian laboratories in Cambridge, Mass., by both the sheriff of Pocahontas and a local farmer. Unfortunately, these objects proved to be not meteoritic. If the Pocahontas Fireball ever reached the ground, it was never found.

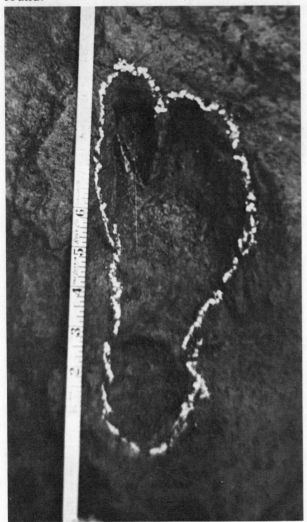

Chinandega footprints, Nicaragua.

49. CERRO NEGRO VOLCANIC ERUPTION

Nicaragua   **October 27**

The Cerro Negro volcano was born in 1850. Since then the volcano has erupted innumer-

able times, usually with frequent lava flows and with the formation of many secondary cones. As a result, Cerro Negro is now really more than one volcano. Yet the main crater, huge and awesome, provided in October a spectacular center ring display of volcanic drama.

The main volcano, located about 20 miles east of the village of Leon, began its 1968 eruptions with intermittent moderate explosions that ejected blocks of fiery incandescent material and produced a dense ash–smoke cloud that extended to altitudes of several thousand feet.

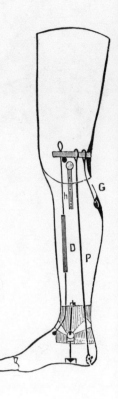

There was also a flank eruption, producing a double vent some 1,000 feet from the Cerro Negro, in a state of nearly continuous activity for 43 days.

During the period from October 24 to December 7, Cerro Negro produced an eruption cloud that reached altitudes of 5,000 to 10,000 feet. The small vent produced a lava flow approximately 6,000 feet long, 300 to 600 feet wide, and 30 to 60 feet high. Intermittent bursts of smoke and ash from the main vent, and a reduced volume of lava from the secondary vent, continued throughout the month of December.

Although the volcano subsided in December, it never really cooled off. The following year, a series of mild gas and ash explosions occurred at the main crater over a ten-day period. The explosions lasted only a minute or two, but each blew a small ash cloud about 2,000 feet high. At first, the explosions came as frequently as four or five times a day, but there were only two on December 28, and one on December 29.

In February 1971, Cerro Negro erupted again, spewing out gases, stone blocks, ash, and other materials and forming a huge cloud that reached an altitude of 20,000 feet. The 1971 eruption was the most violent in recent times, because then no other secondary vents or craters were opened, thus retaining all the volcanic energy in the main crater.

Cerro Negro is the most active volcano in Central America. Since its inception it has displayed many eruptions, usually with frequent lava overflows and the appearance of adventitious cones. Until the 1971 eruptions, the height of the volcano was 1,900 feet above sea level and 700 feet above the surrounding plain. The materials of the volcano are primarily olivinic basalt and some xenoliths of the gabbroid type, as well as of the acid pumiceous type.

Cerro Negro volcanic eruption, Nicaragua.

50. ALLEGHENY RIVER OIL SPILL/ FISH KILL

Pennsylvania   **October 29**

A foamy mass of industrial oil waste slithered down the Allegheny River south of Kittanning, Pa., on October 29. The oil, estimated to be in the thousands of gallons, killed massive numbers of fish. Some 3,000 more gallons of the ooze seeped into the south branch of Bear Creek, causing even more damage. The disaster was investigated by the Pennsylvania State Fish Commission.

51. HAWAIIAN MONK SEAL DISAPPEARANCE

Midway Atoll   **November**

The world population of monk seals has diminished drastically since 1950, so that the species is now nearly extinct. The destroying factor seems to be the mere presence of man. For example, the seal population on the islands of Kure and Green in the Midway chain has been reduced by 50 percent because the islands, once deserted, have now been inhabited. The very presence of people seems to have an inexplicably disturbing effect on seal young.

Seals born in rookeries totally isolated from human society seem to fare much better than those born in breeding grounds near human settlements. Although their ranks are diminishing at an extraordinary rate, the seals have not migrated to other breeding grounds farther away from man. Instead, they remain at their established rookeries and slowly die off.

52. SOUTHERN ILLINOIS EARTHQUAKE

Illinois   **November 9**

Tremors were reported in Minnesota, Arkansas, Illinois, Ohio, Tennessee, Missouri, Kansas, Georgia, Kentucky, West Virginia, North Carolina, South Carolina, Indiana, Nebraska, Iowa, Alabama, Mississippi, Wisconsin, Michigan, and Oklahoma. Although not a major earthquake (Richter magnitude of 5.5), for Americans so used to hearing of earthquakes in remote regions of the world, it provided a curious and frightening backyard jolt. The epicenter was located 120 miles east of St. Louis, Mo., near the Indiana state line.

53. *HESS HUSTLER* OIL SPILL
Delaware   **November 13**

On November 13, a grounded barge leaked oil and stained the beaches of Delaware, doing much harm to the ecology as well as the physical environment. The vessel was carrying 1.8 million gallons of oil. A subsequent investigation was conducted by the Delaware River and Air Resources Commission.

54. ALANDROAL METEOR
Portugal   **November 14**

It was nearly dark when a Portuguese man taking a walk heard a loud noise like a sonic boom. Suddenly, an eerie glow lit up the landscape as though it were daytime. Seconds later, a 55-pound meteorite slammed into the ground about 100 feet away.

The meteor had traveled from southwest to northeast across Portugal descending at a 50- to 60-degree angle. When it struck near the Portuguese village of Jurmenha, it caused a "rain of sand." It also formed a depression in the ground some three inches deep upon impact.

Ten days after the Center received its first notification of the meteorite fall, a specimen weighing 128 grams was cut from the main mass and handcarried from Lisbon to Boston via TWA pilot courier. This specimen was, in turn, cut into several pieces for experiments conducted at various U.S. laboratories.

55. TURKEY EARTHQUAKE
Turkey   **November 14**

A disastrous earthquake of Richter magnitude 6.6 struck northwestern Turkey on November 14.

About 1,000 houses collapsed completely and another 1,300 houses were damaged. A total of 22 people were killed and more than 250 were injured. The quake was associated with a fault-break in the earth about four miles long.

56. MOUNT TRIDENT VOLCANIC ACTIVITY
Alaska   **November**

A ranger at the Katmai National Monument at King Salmon, Alaska, was flying a low-level reconnaissance mission when he noticed that a light ash cover extended like a carpet to the northwest of Mt. Trident volcano.

A further investigation of the ash covering the snow revealed it originated from the volcano and extended for a distance of approximately 60 miles with a fan width of about three miles at its termini.

It was almost as though the volcano had erupted secretly. Some medium-sized bombs were spotted in an estimated half-mile radius around the cone, but there was no sign of a lava flow.

The Center, notified that there had been some activity at the volcano, cabled the Katmai National Monument personnel and obtained dated photographs of the previous weeks that showed there had been no volcanic activity during a two-week period before November 21.

57. SOUTHERN SARGASSO SEA MARINE FOULING
Mid-Atlantic Ocean   **December 5**

The Sargasso Sea is really an island of floating vegetation in the Atlantic Ocean. On December 5, 1968, the cruise ship *R. V. Chain* went out to collect marine surface organisms from the area. Instead of living creatures, the crew found quantities of oil-tar lumps about three inches in diameter in their nets. The nets were so fouled with oil and tar that the *Chain*'s operation had to be discontinued.

58. SOUTHWEST ICELAND EARTHQUAKE
Iceland   **December 5**

A sharp earthquake, registering more than 100 aftershocks during a 24-hour period, shook southwestern portions of Iceland on December 5. The epicenter of the quake was located in a volcanic area about 20 miles south-southwest of the capital city, Reykjavik.

59. ITHACA FIREBALL
Ithaca, New York   **December 9**

A bright fireball spouting blue flames passed over south-central and southeastern New York State on December 9. Many witnesses claimed to have seen the blue flame pass over houses and ultimately crash "with a loud noise that sounded like dynamite." Explosions were reported to have shaken several houses.

Although several dozen airborne and ground-based reports were received by the Center, and a photograph of the fireball taken by an astronomer at the Springhill Meteor Observatory gave an excellent indication of the meteor's

trajectory, no search was undertaken due to the lack of sufficient information needed to determine an accurate impact point.

60. OIL SPILL
Panama   **December 13**

The oil tanker *Witwater* broke up in Panamanian territorial waters on the Atlantic side of the Panama Canal on December 13, spilling 14,000 barrels of oil and threatening marine flora and fauna in the tropical area.

Because the spill was only five miles from the Smithsonian Institution Tropical Research Institute in Panama, daily field trips were made to the area by Smithsonian personnel with the cooperation of the U.S. Coast Guard. Also, Kenneth Biglane of the Federal Water Pollution Control Administration flew to Panama to assist in programs of containment and removal.

Several thousand barrels of oil were successfully removed from causeway waters by pumping. Another approximately 5,000 barrels were ignited by thermite grenades which reduced the oil thickness in a 4,000-square-foot pool from four inches to less than an inch.

By December 17, it was estimated that more than half of the spilled oil had been removed. However, 8,000 barrels of oil still remained in the *Witwater's* bow section and the bow was leaking oil at the rate of 50 barrels a day.

The marine life in the area suffered from the oil discharge, with a heavy loss of crabs and other tidewater and shore creatures.

61. ARIZONA FIREBALL
Arizona   **December 14**

On December 14, an extremely bright meteor traveled northwest to southeast across Arizona and was observed from cities as widely separated as Flagstaff and Holbrook.

62. SANTA BARBARA EARTHQUAKE SWARM
Santa Barbara, California   **December 28**

The December 28 earthquake did not "rock" Santa Barbara, it only jiggled the area a bit. The quake, which registered Richter magnitude 3 to 3.5, was just one rumbling in a swarm of earthquakes felt in the Santa Barbara Channel since June.

63. WAKULLA OIL SLICK
Florida   **December 30**

Wildlife and marine life along the coastal area of Wakulla County, Florida, from St. Marks to Piney Island, were threatened by a crude oil spill from a barge owned by the Seminole Asphalt Company of St. Marks.

A survey on the oil slick's effects on marine life showed that many types of sea life were dying—or were already dead—from hydrocarbon poisoning and lack of oxygen. Many ducks, snipe, and other birds were so covered with oil that they were unable to fly. Smaller birds faltered while trying to walk in the heavy oil. An investigation of the spill was undertaken by the U.S. Coast Guard.

Cerro Negro volcanic eruption, Nicaragua.

EVE.NT REPORTS   1969

22   Akuri stone-axe tribe discovery, Surinam, South America. *Photo courtesy Art and Evy Yohner, West Indies Mission.*

1. AKURI STONE-AXE TRIBE DISCOVERY

Surinam, South America   1968–1969

They were a tribe of nomads who roamed the Surinam jungle in constant search of one of their main diet staples—honey—and they had never seen a civilized man. Nor had a civilized man ever seen them. Missionaries of the West Indies Mission had been looking for this stone-age tribe for three years and, in June 1968, they made their first contact.

The tribe was not exactly hospitable to the missionary expedition, but the missionaries' presence was tolerated by the Akuriyo—or "Akuri people"—long enough for some basic observations to be made.

Akuri stone-axe tribe discovery, Surinam, South America. *Photo courtesy Art and Evy Yohner, West Indies Mission.*

Although the tribe depended on the stone axe as their most important tool, they possessed other primitive implements, such as bamboo knives used for cutting hair and skinning animals. They also made other tools by embedding a rodent's tooth in the end of a small hardwood stick, creating a kind of miniature pickaxe used for gouging and making holes. (Rodents apparently play a major role in the Akuriyo's life; in fact, the word "akuri" means "agouti," a small rodent of the South American jungles.)

Tooth and seed necklaces are worn by both men and women. Otherwise the women wore nothing more than a brief apron made of seeds, and the men wear only a narrow loincloth woven from palm fibers.

The Akuriyo are skilled hunters, relying on bows and curare-tipped bamboo arrows. They are strictly nomads who apparently roam over some 6,000 square miles of rainforest. The missionaries discovered that one of the main reasons for this restlessness is the continual search for honey, which they apparently cherish, but of which they never can find enough. The other main components of their diet, kwai fruit, meat, and the thorn palm nut, are found in abundance.

The missionaries' first two contacts were made with the same group of 28 Akuriyo, although two more were present on the second visit. In February, a third visit was made, and the missionaries discovered that the tribe had now divided itself into two temporary villages. In the larger village, the Akuriyo continued to be hostile to the missionaries and government officials in the expedition party, but not to the group of Trio Indian guides.

On the fourth excursion in June, the missionaries made contact with an entirely different group of the Akuriyo. This group of 25 Akuri tribesmen was friendly and receptive and seemed less afraid than those encountered on the previous expeditions.

The missionaries introduced these friendly tribesmen to the smaller wonders of modern

Akuri stone-axe tribe discovery, Surinam, South America. *Photo courtesy Art and Evy Yohner, West Indies Mission.*

civilization. Among the gifts presented to the Akuriyo by the missionaries were steel axes, machetes, knives, and files.

The Surinam government also began steps to establish close contact with the Akuri groups, while at the same time, protecting them from external exploitation and disease—the unwanted elements of modern civilization.

2. MOUNT RAINIER SEISMIC ACTIVITY
Washington State   1968-1969

The seismology station operated by the University of Washington reported that seismic activity at Mt. Rainier had been increasing steadily since 1968. Between April and June 1968, there were one to three tremors recorded every five days. In September, the rate increased to about one a day.

By April 1969, the tremors still occurred once a day. In May, the activity increased again; and, beginning June 15, tremors were recorded at nearly two per day.

3. CHIRIMACHAS ADVANCE
Peru   1968-1969

The Chirimacha, an insect that carries a painful, fatal illness known as Chagas' disease, proliferated throughout a large area of southern Peru in 1968 and 1969. The insect, *Triatoma infestans*, is resistant to common pesticides and is extremely difficult to control. The disease, actually caused by a parasitic protozoan, *Trypanosoma cruzi*, nearly always is deadly and is characterized by intense headaches, nausea, fainting, nosebleeds, and finally a pronounced lethargy. There is at present no cure for or vaccine against the disease.

The insect is very much like the ordinary housefly in its habits, but is four times as large. Although the insect thrives on garbage and prefers living in warm damp spots, it can also live for months without food or air. The chirimacha also reproduces at an extraordinary rate.

The Peruvian government made several attempts to eradicate the deadly pest in affected neighborhoods, but none were successful.

In despair, many families abandoned their homes to the chirimacha and moved to surrounding mountain areas which are relatively free of the insects. The people who remained, however, continued to fight a losing battle against the chirimachas' advance.

4. TOLEDO CATTLE KILL
Toledo, Spain   1968-1969

For more than a year cattle in the province of Toledo, Spain, were dying at an alarming rate from an unknown cause. Investigations later revealed that the deaths were caused by the ingestion of a poisonous plant, *Denante crocata*. Immediate steps were taken to exterminate the toxic plant.

5. AMAZON RIVER PORPOISE DISAPPEARANCE
Peru   1968-1969

The Lima, Peru, newspaper *El Comercio* reported that a species of porpoise indigenous to the Amazon River was in danger of extinction because of industrial exploitation. The newspaper reported that the Peruvian Ministry of Agriculture had prohibited the capture of these mammals. However, the Ministry later issued a statement denying it had ever known of the impending disappearance of the porpoise. Moreover, the Ministry declared that the capture of the animal was not prohibited.

6. SOLOMON ISLANDS EARTHQUAKE
Solomon Islands   January 5

The first large earthquake of the new year was magnitude 7.3 Richter, with its epicenter located in a submarine fault zone north of and parallel to the island of Santa Isabel, one of the major islands in the Solomon chain. Powerful surface waves created by the quake were recorded up to 150 miles from the quake's center.

7. SANTA FE PREHISTORIC CAMEL TRACKS
Santa Fe, New Mexico   January

Camels seem almost synonymous with the Sahara, but fossil remains indicate that the animal once lived in North America. The theory was reinforced when workmen found prehistoric camel prints in a New Mexico stone quarry.

The set of regular depressions were discovered in a layer of soft stone under a cinder bank. The quarry is 12 miles southeast of Santa Fe, the home of the Museum of New Mexico. Museum personnel investigated the tracks and determined that they were petrified camel tracks some 70,000 to 150,000 years old.

After the announcement, the Federal Bureau of Land Management withdrew the site from public use and built a shed over the tracks to preserve them from weather erosion.

8. OITA-TOKYO FIREBALL
Ise Bay-Oshima Island, Japan   January 7

A pale-blue fireball widely observed over a large part of Japan in fact may have been part of a rocket. The object was reported traveling north at a high rate of speed, leaving a smoke trail behind as it disintegrated in the atmosphere. Some observers claimed that the fireball split into three parts in its descent.

The fireball, however, might have been part of the booster rocket associated with the Russian satellite Cosmos 261. The rocket body had been predicted to re-enter the earth's atmosphere at about the same time as the fireball was seen. Five visual sightings in Manitoba, Canada, confirmed the satellite re-entry on January 7 and the satellite track did move over the Japanese islands at about the same time the "Oita-Tokyo Fireball" was sighted.

9. MOUNT MERAPI VOLCANIC ERUPTION
Java, Indonesia   January 8

The 7,800-foot-high Mt. Merapi on the island of Java has had a history of intense volcanic activity since October 1967 when its dome began to grow from the extrusion of lava. In early January 1969, the volcano erupted with new fury, sending *nuée ardente* avalanches four miles down the slope and shooting ash clouds over 5,000 feet into the sky.

The volcano continued its spectacular eruption the next day, when strong quakes were recorded accompanied by *nuée ardente* explosions and lightning. In fact, the explosions were of such a magnitude they could be heard in the city of Jogjakarta, some 25 miles south of the volcano.

Later in the day, huge reddish *nuée ardente* clouds covered the volcano's peak. When these clouds cleared and the lightning and the rumblings ceased, it appeared to observers that the entire lava dome created by the 1967–1968 lava flows had slid down the mountain slope.

Between January 13 and January 20, the activity of Mt. Merapi included incandescent avalanches occurring on the average of every ten minutes. The month-long lava flow around the collapsed 1967–1968 lava cone formed a mass some 1,200 feet long and some 400 feet wide.

10. ATHENS FIREBALL
Athens, Greece   January 16

A suspected satellite re-entry was observed over the Athens area when an object with a fireball appearance was seen traveling from west to east. The object seemed to be moving relatively slowly, left shockwave patterns, and split in two main pieces before disintegrating completely.

11. MALAWI FIREBALL
Malawi, East Africa   January 21

The meteor appeared as a glowing red and yellow fireball that crossed the African skies in five to ten seconds and landed in the bush country of Malawi. The fireball left a train described by one observer as being something like a magnesium flare. It was extremely bright, producing enough light for some observers to read their wrist watches. About one or two minutes after the meteor had vanished and left the night sky black again, four explosions were heard, followed by a rumbling sound like distant gunfire. According to some reports, the rumbling lasted as long as a minute and was strong enough to make windows vibrate.

The meteor landed with an impact great enough to register a reading on a seismograph at Chileka, about 25 miles away from the impact point. The data recorded on the instrument coincided with visual observations on the meteor landing. Unfortunately, the seismographic readings could place the impact point only within a general area; thus, locating the assumed meteorite became a difficult problem. A search was begun, but because the position of the impact crater was not certain and the terrain is rough and marshy and covered by dense tropical bush, the search was not successful.

Scientists are fairly sure that the seismographic reading was caused by a genuine meteor. No satellites of any comparable size re-entered the earth's atmosphere at that time.

12. ICELAND POLAR BEAR KILL
Grimsey, North of Iceland   January 22

The death of a polar bear on Grimsey was related to the movement of arctic pack-ice near that island. It was the third consecutive year that the ice had moved south into the waters around western and northern Iceland.

13. ST. MARY'S DUCK KILL
St. Mary's, Maryland   **January 23**

Approximately 300 ducks, mostly scaup, were killed when they crashed into buildings on the campus of St. Mary's College. Subsequent investigations led to the conclusion that a combination of very heavy fog and light from newly installed mercury vapor street lamps caused the ducks to mistake the parking lots for water and to inadvertently strike the buildings.

14. CEDAR RAPIDS FIREBALL
Cedar Rapids, Iowa   **January 26**

On the morning of January 26, a very bright fireball was observed over Iowa and Minnesota. A report from as far away as Lansing, Mich., said the fireball lit up the sky like "sheet lightning," but the object was visible for only ten seconds, and no accompanying acoustical phenomenon was reported. The end-point of the fireball was calculated to be somewhere in central Wisconsin.

15. GUNUNG IJA VOLCANIC ERUPTION
Flores Island, Indonesia   **January 27**

The Gunung Ija volcano erupted for about a month, ejecting smoke and lava and causing explosions and volcanic earthquakes. The first eruption created a three-mile-high pillar of smoke in shades of black, white, yellow, and blue. The entire mountain from top to lower slope burned for some three hours, and a loud thundering sound continued for almost seven hours before it stopped abruptly "like the engine of an automobile being stopped."

16. SANTA BARBARA OIL SPILL
California   **January 28**

On January 28 an oil well five and a half miles off the Santa Barbara coast began spewing millions of gallons of oil over hundreds of square miles of water and shore. The massive oil spill began when the drill on Union Oil Company's Platform A suddenly struck a gas deposit 3,000 feet below the ocean floor. The gas, under enormous pressure, forced its way up through the drilling hole, bringing crude oil with it. The oil and gas continued to gush from the well until the flow was temporarily controlled by plugging the well with cement.

Before the well was plugged, however, an area of the Santa Barbara Channel, estimated by one news agency as over 800 square miles, was covered with a thick layer of oil.

Santa Barbara oil spill, Santa Barbara Channel, California. Baby California sea lions, coated with crude oil, on oil-soaked beach on San Miguel Island, 45 miles from Santa Barbara. *Photo courtesy of Richard Smith, Santa Barbara News-Press*

Exactly how much oil was spilled varies with the source, and estimates of the average daily flow from the well during the initial blowout differed by factors of nearly 100. For example, one report estimated the leakage at 300 to 500 barrels a day. Another supposedly "conservative estimate," based on the area and thickness of the oil slick, said 5,000 barrels a day. Still another report placed the figure at 16,000 barrels a day. One news agency even quoted a leakage rate of 21,000 barrels a day.

After the drilling hole had been cemented, oil seepage continued at a very slow rate. Then, on February 12, new and less easily controlled oil leaks were created when gas under pressure created several fissures in the ocean floor and through the shallow oil sands. The wild uncontrollable leakage continued at a rate of 500 barrels a day through late February. During the first week of March, however, the amount of leaking oil was reduced by at least half.

One so-called "conservative" estimate claimed that over three million gallons of oil had spilled into Santa Barbara Channel by the middle of May. In July, the spill was reduced considerably both by pumping oil out of the underwater reservoir and by improved oil collection techniques. A relatively low level of leakage continued until December 16, when a new oil slick appeared over a 50-square-mile area.

On December 17, the Union Oil Company reported a leak from a cracked pipe under Platform A. Pumping was shut down long enough to repair the pipe, but when the pumping was stopped, massive amounts of oil were again pushed up through the ocean floor. Attempts to contain or stop the leakage proved only partially successful and oil continued to pollute the water and shores of the Santa Barbara Channel.

Since the initial blowout, about 100 miles of California coastline had been contaminated with oil. The December 16 leak carried new oil pollution to a 12-mile stretch of beach southeast of Santa Barbara in an area where vast numbers of migratory birds spend the winter.

Except for the damage to sea and shore birds, the overall effects of the oil pollution on wildlife in the Santa Barbara Channel are still largely unknown. One scientist remarked a month after the blowout, "the only absolute evidence we have all seen with our own eyes is the effect on the birds. There are 20 to 50 dead birds that have come ashore. We don't know how many dead are out at sea. The birds all die if they settle in the oil, for they attempt to clean it off their feathers by preening, thus they naturally ingest the oil which is toxic."

The Santa Barbara *News-Press* reported that estimates of dead birds ranged from 3,000 to 15,000 (and higher). One report claimed that only 10 to 15 birds out of some 1,500 birds treated for oil damage were let loose—and it was unknown whether they survived. Sea birds that did survive or were not contaminated by the initial leakage returned to the area to winter. Of these, one flock of about 200 grebes became contaminated by the oil.

Santa Barbara oil spill, Santa Barbara Channel, California. California murre completely immobilized by oil waits for death on an oil-soaked beach. *Photograph courtesy of Richard Smith, Santa Barbara News-Press*

Other species of birds were also contaminated: cormorants, loons, common grebes, eared grebes, California murres, scoters, marbled godwits, willits, dowitchers, western gulls, California gulls and Heerman gulls.

It was even more difficult to assess damage to the marine environment of Santa Barbara Channel.

Certain populations of marine invertebrates seemed to have been killed outright by the oil, while other species appeared to suffer no ill effects whatsoever. One scientist reported that the oil was causing serious damage in the intertidal zone, expecially to abalone, mussels, and vegetation. Also, large numbers of periwinkles and "wavy tops" were found dead; yet, oddly enough, most of the sea anemones remained alive. The tidepools with oil were also reported to be empty of nudibranches and rock crabs. On beaches at Point Conception, sea anemones and mussel colonies were seemingly normal and healthy in spite of heavy oil coating.

Fish eggs and larvae in areas of oil contamination appeared to be normal. Although surface phytoplankton and microplankton were reduced in the months after the spill, there was no convincing evidence that the reduction was due to the oil itself. The leakage also did not seem to have an observable influence on fish life in the area.

A few sea mammals were found dead, but, for the most part, they seemed to avoid any direct contact with the oil. An unusual number of California gray whales, six, were found dead on the beaches, but there again was no direct evidence linking the deaths to the oil disaster. One bottlenose dolphin was found suffocated when oil clogged its blowhole, but most reports indicated that the fur seals, elephant seals, and California sea lions on the islands in the Santa Barbara Channel were not harmed by the oil.

It was also reported that Corexit, the chemical most widely used as an oil-dispersing agent (causing the oil film to break up into small droplets which then sink to the bottom) was also toxic to many forms of marine life. So far, however, little evidence has been found to link Corexit with damage to sea life.

17. MINDANAO EARTHQUAKE
Philippine Islands   **January 30**

A strong earthquake, registering a Richter magnitude of 7.2, shook the Philippine islands of Mindanao and Samar. The quake's epicenter was located about 85 miles out at sea. On February 3, there were two subsequent aftershocks, the first at Richter magnitude of 6.0 and the second recorded at magnitude 6.8.

18. MAGHAR LANDSLIDE
Israel   **January**

An intensive landslide caused great damage in the Israeli village of Maghar located on the steep southern slope of Mt. Hazon. The landslide, which came in sporadic waves over a period of five days, was shallow, interrupting only the soil cover and the upper marly strata on the slope. The village had been damaged by landslides in 1899 and 1928.

19. HOUSTON FIREBALL
Houston, Texas   **February 6**

A bright meteor lit up the sky over the Houston area and a loud sonic boom rocked the city seconds after the sighting. The acoustic phenomenon was recorded by the National Aeronautics and Space Administration seismograph in Houston, but no meteoritic material was found.

A second fireball was seen over the Houston area on May 3.

Santa Barbara oil spill, Santa Barbara Channel, California. Workers picking up the tons of straw that soaked up oil in Santa Barbara harbor.
Photo courtesy Richard Smith. Santa Barbara News-Press

20. PUEBLITO DE ALLENDE METEORITE SHOWER
Hidalgo del Parral, Mexico   February 8

On the evening of February 8, a blue-white fireball turned night into day over a 1,000-mile path from central Mexico to southern Texas and covered a large area of rural Chihuahua State with at least four tons of meteoritic material. This shower of meteorites produced the largest number of carbonaceous chondrite stones ever recorded. And, of the four tons of material estimated to have reached earth, at least two tons were collected for analysis.

Thousands of individual stones rained down over a wide area surrounding the village of Pueblito de Allende, south of Chihuahua on the Pan-American Highway. The largest single specimen found weighed 110 kilograms, but was broken into many pieces on impact.

The meteor that produced the shower was spectacular. Hundreds of observers throughout Mexico reported seeing the brilliant flash of light and hearing a tremendous explosion. But the most spectacular phenomena were reported around the city of Hidalgo del Parral in the south-central part of Chihuahua State, where the fireball was accompanied by detonations and a strong air blast.

Pueblito de Allende meteorite shower, Hidalgo del Parral, Mexico.

Air Force meteorologists flew a B-57 through what they calculated might be the dust trail of the shower and collected samples of atmospheric dust with special filter traps aboard the plane.

The Allende meteorites were positively identified by both Smithsonian and NASA scientists as carbonaceous chondrites—very different stones from the more common iron-nickel type of meteorite. These meteorites contained chondrules, or clusters of small, beadlike objects rich in magnesium and containing significant amounts of olivine and some glass. Some stones also contained grossular and sodalite, two minerals not previously found in meteorites.

The shower was a great boon to meteoriticists, and within only weeks after the shower, the Smithsonian had distributed Allende material to 37 laboratories in 13 countries.

21. TELICA VOLCANIC ACTIVITY
Leon, Nicaragua   February 11

The Telica Volcano, located 12 miles north of the city of Leon, began a phase of relatively mild activity characterized by light ashfall and steam emission on February 11. The peak of the activity occurred between February 14 and 18.

By the end of the month, the activity had apparently ceased. But, on May 14, activity resumed with ash and fine lapilli, or volcanic stones, being erupted. These eruptions stopped on May 17. Talica exhibited some mild activity during the summer, particularly in mid-August and mid-September, but then became quiet once again. So quiet, in fact, that a small lake formed at the bottom of the crater by the end of the year.

22. ANCHORAGE MOOSE MIGRATION
Anchorage, Alaska   February

Unusually large numbers of moose began a migration into populated areas of Anchorage, their movement apparently triggered by a deep snow that forced the animals to seek out more accessible feeding areas.

The migration was also caused in part by the encroachment of humans on the moose range through the spread of suburban developments. An increase in the moose population over the past few years also contributed to the confrontation between man and moose.

23. DECEPTION ISLAND VOLCANIC ERUPTION
Antarctica   February 21

Volcanic activity on Deception Island, characterized by the ejection of ash, pumice, and volcanic bombs, was followed by an earthquake and mudslide that destroyed a nearby Chilean research center and opened a landlocked lake at the sea.

There were two other research stations in the area, one British and the other Argentine. The Argentine station suffered little damage, but the British station was heavily damaged by mudflow. A graveyard and most of a whaling station in the vicinity were also destroyed. Kroner Lake, which had been landlocked, was opened to the sea in the turmoil.

The mudflow in the area known as Penfold Point was four to seven feet deep. In another area, called Whalers Bay, the flow was quite liquid, and moved very rapidly—at speeds up to 20 to 30 miles an hour. Beaches were heavily littered with timber, food cans, fuel drums, and at least one grave marker, most of the debris carried there by flows from the Whalers Bay area.

The flow carried large blocks of glacier ice weighing several tons to the coast. Similar blocks of glacier ice were reported at and near the Chilean station and a few very large blocks at the British station.

The Chilean station was completely destroyed by a combination of ashfall and mudflow several inches deep. Large volcanic bombs struck nearby and water vapor and hydrogen sulfide gas were detected, while elemental sulphur had been deposited from numerous small fumaroles and vents. The ground temperature on the hillside behind the Chilean station fluctuated between 50 and 70 degrees centigrade.

24. MACASSAR STRAIT EARTHQUAKE
Indonesia   February 23

Sixty-four people were killed and 97 injured in a destructive earthquake near Madjene, Celebes Island, Indonesia. The damage was extensive, with 1,287 structures in three localities being destroyed. The quake was recorded at a Richter magnitude of 6.9, although severe seismic seawaves generated by the quake amplified its destructiveness and damaged coastal villages on the island north of the city of Madjene. The quake's epicenter was located in the Macassar Strait.

25. KILAUEA VOLCANIC ERUPTION
Island of Hawaii   February 24

The Kilauea volcano had erupted in August and October 1968, but that activity was mild compared with the activity that began in February and continued for more than a year.

In the eastern rift zone of the volcano, there had been four eruptions in a 15-month period.

Another eruption along a fissure between the volcano craters, Alae and Aloi, began on May 24. Between then and the last activity October 20, there were 11 phases of high lava fountaining and significant lava output.

The phases were spaced from two days to four weeks apart. Fountains from 100 to 1,500 feet high issued from a 300-foot-long segment of the fissure. On the downwind side of the fissure a cone of welded spatter at least 100 feet high had been built.

The most eruptive phases lasted less than 10 hours, with fountain activity building gradually and then ending rather abruptly, usually in a matter of a few minutes. Flows from the 11 phases completely filled Alae crater and partially filled Aloi. One lava flow, from the fourth phase on June 25, actually reached the ocean some seven miles from the vent.

Between the phases, the lava column could generally be seen in the fissure at a depth of several score feet. The column usually rose slowly within the fissure without gas emission. Then, quite suddenly, bubbling began, and the column dropped rapidly with violent spattering. The cyclic rise and fall normally occurred every 10 to 15 minutes, with the actual fall lasting only about two to five minutes. Occasionally the rising lava column reached the surface and created a small flow, generally confined to the vent area.

Kilauea volcanic eruption, island of Hawaii. Forty-foot-high dome fountain on east rift of Kilauea. *Photo courtesy of U. S. Geological Survey, Washington, D.C.*

Also reported were significant ground deformations that accompanied the phase and interphase period. The summit of Kilauea, four miles from the vent area, swelled upward and outward between the eruptive phases and contracted rapidly when there was high fountaining and a great deal of lava output. On the other hand, the east rift zone near the eruptive site generally expanded during phases and contracted during interphases.

The eruption that began on May 24, 1969, between the Aloi and Alae craters on the upper rift zone of Kilauea continued into 1970. The continuous activity became the longest and probably the most voluminous of all historic Kilauea flank eruptions. The mountain has changed in the process, and a new shield on the northeast side was named Mauna Ulu, which means "growing mountain."

Kilauea volcanic eruption. island of Hawaii. Tract of highway severely damaged by the volcano. (Note center dividing line.)

A fissure cut the southwest flank of Mauna Ulu on April 9 and it spread westward across Aloi crater and some 2,500 feet beyond. Lava erupting the fissure filled the crater, but activity died out by May 1. The fissure on the northeast flank of Mauna Ulu was active periodically in the fall and winter of 1969, but it remained peaceful in 1970, until July 6, when low fountains burst from it. Since then, eruptive activity from the fissure had been more or less constant.

There had also been activity in other Kilauea fissures in 1969 and 1970.

The summit of Kilauea expanded greatly from December 1969, to June 1970, a period when eruptive activity at Mauna Ulu was at a comparatively low ebb. Similarly, there had been little summit formation since the June activity.

26. MOUNT EBU LOBO VOLCANIC ERUPTION
Flores Island, Indonesia   February 27

Mount Ebu Lobo, Flores Island, Indonesia, began erupting on February 27; the volcanic activity was characterized by the emission of fire, steam, and ash.

27. PORTUGUESE EARTHQUAKE
Portugal   February 28

The strongest earthquake recorded in the world since the Alaska rumble of 1964 struck Morocco, Spain, and Portugal on February 28, causing at least three deaths and many injuries. The epicenter of the quake, which registered over 8.0 magnitude on the Richter scale, was located in the sea 145 miles west of Cape Vincente, Portugal. Ships in the epicentral region were badly shaken.

The earthquake produced a tsunami reported in Casablanca as seven feet high. Along the Portuguese coast, the wave was recorded as three feet high with a mean velocity of 240 miles per hour.

Almost two hours after the main tremor was felt, a major aftershock of 6.2 magnitude was recorded, followed by many smaller tremors. In all, some 367 aftershocks were recorded, with the last on May 5. Fissuring also occurred near Algarve, Portugal, but there were no landslides reported.

The quake was so strong that it was measured very clearly at a seismographic station in Boulder, Colo.

28. DUTCH COAST OIL SPILL
North Sea   February

Fuel oil in large quantities polluted North Sea waters, contaminating beaches and oiling birds off the Netherlands coast. The exact source of the spill is unknown, but most likely was the result of oil dumped from a ship.

The Dutch Society for Animal Protection reported that 15,000 to 25,000 oiled seabirds came ashore and that the probable total kill approached 100,000.

In the last weeks of February, birds dead from oil pollution began to appear on the British coast. It is believed that the birds, mostly auks and puffins, were casualties of the same spill that ravaged the Netherlands coastline.

29. PERUVIAN ANT IRRUPTION
Peru   **March**

Making a rustling sound resembling burning brush, millions of leaf-cutting ants, *Atta sexdens*, marched across farmland in the valleys of La Convención and Lares in Peru. The army of ants (not a flesh-eating species) caused widespread devastation of cultivated crops in southern Peru, especially of the tea and cocoa crops.

Farmers in Peru and other South American countries have faced ant irruptions in the past, but the 1969 invasion was larger, and therefore more destructive and uncontrollable, than any in recent memory.

Contact poisons such as Aldrin and Dieldrin were used, but best results were with poisonous powders such as Mirex. However, even poison could not control the vast number of insects. It was the ants themselves who ended their march. By heading for the open lands after ravaging the cultivated areas, they became victims of their own numbers. The countryside could not support their vast population.

30. COOK INLET OIL SPILL
Cook Inlet, Alaska   **March 4**

The S.S. *Yukon*, carrying 150,000 barrels of oil, ruptured its hull on an unidentified submerged object and spilled an unknown amount of oil into Cook Inlet, Alaska. An investigation of the incident was sponsored by the Multiagency Screening Group. The inlet suffered another spill later in the year.

31. TECOLUTLA SEA ANIMAL DISCOVERY
Tecolutla, Mexico   **March 6**

The initial reports of the discovery of a possible prehistoric monster sounded astounding.

The creature, washed ashore on the beach at Tecolutla, Mexico, was reported as a mammal, measuring 30 feet long, weighing about 30 tons, and sporting a single 30-inch horn. It was something man had never seen before, only imagined in science-fiction stories.

The carcass, upon scientific investigation, however, proved to be only a rather common whale. The animal apparently had been attacked by sharks and its skull had been fractured and split. That long, mysterious horn described in the initial and, sadly, erroneous report was nothing more than an extended piece of bone.

32. MINAMI-DAKE VOLCANO
Kyushu, Japan   **March 8**

Characterized by voluminous ash eruption, Minami-Dake exploded into activity on March 8. The volcano is located on the Kagoshima Prefecture in southern Kyushu. Reports stated that smoke from the eruption rose to a height of nearly 5,000 feet.

33. FARALLON DE PAJAROS SUBMARINE VOLCANIC ERUPTION
Northern Marianas Islands   **March 11**

Steady acoustic activity near Farallon de Pajaros in the northern Marianas Islands in the Pacific Ocean was recorded by United States government stations on Eniwetok, Wake, and Midway islands. A fishing boat sailing near Farallon de Pajaros in the Marianas reported a submarine eruption on the same day. Fishermen said they heard three explosions and noticed that the neighboring sea surface had turned brown in color.

34. URKUT LANDSLIDE
Mount Bocshoc, Hungary   **March 14**

For five days, from March 14 to 19, a landslide destroyed 250 feet of highway between the Hungarian towns of Ajka and Urkut. The slide was presumably caused by heavy rains in the Mt. Bocshoc region.

35. LOUISIANA OIL SPILL
Gulf of Mexico   **March 16**

High seas shifted an oil rig in the Gulf of Mexico located seven miles off the Louisiana coast causing the control valve on a recently completed well to be torn away. The well had been drilled to a level of 11,400 feet when the valve was torn off, and high gas pressure caused the oil to surface.

The amount of the oil that leaked out was unknown, but it accumulated into several large, well-defined slicks on the water's surface. Northerly winds, however, blew the oil away from the coastal area. Within a few days, the slicks were dissipated by high seas. There was no evidence of damage to marine life; and, on March 19, the oil well was capped.

36. CHICAGO OIL SPILL
Chicago, Illinois   March 21

A Procter and Gamble Company facility accidentally ejected some 5,000 gallons of soybean oil into the Chicago River. At the site of the spill, the percentage of dissolved oxygen—necessary for marine life respiration—was zero. The prime effect of the oil spill was to lower the dissolved oxygen content of the water farther down river.

37. DIDICAS VOLCANO
Babuyan Island, Philippines   March 27

The Didicas Volcano remained a submarine volcano until 1952, when it erupted to form an island 1.5 miles in diameter and about 800 feet high.

On March 27, new activity at Didicas was confirmed by an air reconnaissance flight. The eruption was characterized by ash and steam emissions; and, on March 31, black smoke was seen gushing from the island.

The spastic ejection of rock originated from a vent opened during the earlier stages of activity. The volcano continued to build up pressure, demonstrated by the steady rise and extrusion of ashes and steam from a crater on the northern side of the islet.

Investigations of the new activity were conducted from the air. Landing on the volcanic island was impossible.

38. HALMAHERA ISLAND EARTHQUAKE
Indonesia   March 27

An earthquake of Richter magnitude 7.0 shook Halmahera Island on March 27, but no serious damage on the island was reported. The quake was actually the seventh in a series of earth tremors during a seven-week period beginning January 30.

39. TURKEY EARTHQUAKE
Turkey   March 28

The largest earthquake in a series of tremors struck western Turkey, demolishing buildings, killing 53 persons and injuring another 100, as well as rendering inoperative all communication lines. The main jolt of the series of quakes

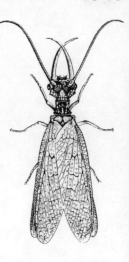

registered magnitude 7.5 on the Richter scale.

The tremors continued into April, with the most powerful occurring on April 1. The aftershocks occurred along the Anatolian fault and hampered the operations of rescue parties attempting to help those injured and homeless.

40. ETHIOPIAN RIFT EARTHQUAKE
Sardo, Ethiopia   March 29

The town of Sardo, Ethiopia, was completely destroyed, leaving 24 dead and 163 injured, when an earthquake struck on March 29. The quake registered a Richter magnitude of 6.2. Its epicenter was located in Ethiopia near the southern tip of the Red Sea.

Ethiopian rift earthquake. Sardo. Ethiopia. Remains of the town of Sardo. *Photo courtesy Prof. Pierre Gouin. Haile Selassie I University.*

Five shocks occurred the same day, ranging in magnitude from 4.0 to 5.7 on the Richter scale. Faults were observed over a six-mile stretch of the Kombolcha-Assab Highway. The largest faults and rockslides occurred on April 5. Aftershocks continued until April 8.

The huge faults and cracks, looking like deep scars in the earth, occurred near the destroyed town of Sardo. The largest displacements occurred about three miles east of the ill-fated town.

The Red Sea quake was the third major earthquake to rock the Middle East in a period of four days. Previous earthquakes struck Turkey on March 28 and Ethiopia on March 29. The disasters were apparently connected: seismicity in the Middle East and East Africa during March and April showed a time-distance relationship extending from western Turkey to the southern end of the African rift system.

Ethiopian rift earthquake, Sardo, Ethiopia. The Kombolcha-Assab Highway following the earthquake. *Photo courtesy Prof. Pierre Gouin. Haile Selassie I University.*

41. RED SEA EARTHQUAKE
Red Sea   **March 31**

Hundreds were left homeless and 65 persons were killed when a strong earthquake of magnitude 6.5 shook the Middle East. The quake originated south of the Sinai Peninsula at the northwest tip of the Red Sea.

The effects of the quakes were felt in Israel, Egypt, Iraq, Lebanon, Turkey, and Italy. Two shocks followed the main tremor on the same and following days, with Richter magnitudes of 5.0 and 4.7, respectively.

The quake was preceded by foreshocks in the same area on March 24, when several dozen tremors were registered at Richter magnitudes of 4.4 to 4.5.

The Red Sea quake was the third major earthquake to rock the Middle East in a period of four days. Previous earthquakes struck Turkey on March 28 and Ethiopia on March 29. The disasters were apparently connected: seismicity in the Middle East and East Africa during March-April showed a time-distance relationship extending from western Turkey to the southern end of the African rift system.

42. GREENVILLE FIREBALL
South Carolina   **March 31**

A greenish-blue fireball was sighted traveling in a southeast to northwest direction over the Piedmont area of South Carolina. The object, with a greenish-colored tail, exploded into many fragments. No sonic phenomenon was reported, and no meteoritic material was discovered.

43. SALMON HIGH-DDT-RESIDUE
         DISCOVERY
Lake Michigan   **March**

Investigators from the U.S. Food and Drug Administration analyzed numbers of a species of salmon indigenous to Lake Michigan. They found DDT residues in the fish up to a dangerous 19 parts per million of water.

The discovery led to the decision to set the first pesticide residue levels for fish sold in interstate commerce. On April 22, Robert H. Finch, secretary of the Department of Health, Education and Welfare, set an interim residue level for DDT in fish at five parts per million, pending further study.

44. COLOMBIAN TRIBE DISCOVERY
Colombia   **March**

An apparently lost tribe of Indians was discovered in the Colombian forest near Cuqueta River, a tributary of the Amazon. The Indians were thought to be members of the Yuri Tribe, once believed extinct.

The discovery was made by a search and rescue party attempting to find Julian Gil, a fur trader, who had apparently made the first contact with the tribe and was killed in the effort.

The tribe appeared to be semi-sedentary horticulturists. Gardens of peach palms and bananas were seen around the Indian village, which consisted of a single *maloca*, a large conical structure housing all 200 members of the tribe. The trees in the gardens were cut in such a way as to suggest the use of a stone axe. (A later expedition revealed that stone axes were indeed the tribe's primary tools.) Evidence of other similar villages in the area was also found.

Early reference books and missionary reports showed that a tribe called the Yuri inhabited the region at the headwater of the Prue River. Three Yuri word lists had been collected in the middle 1800s, and linguists maintain that the language is not related to any other.

There were three expeditions to the Yuri area. The first was made by Julian Gil's brother and a detachment of marines from the Colombian navy. The expedition reached the Indian village where they found signs of Julian Gil's presence—a necklace made from the buttons from Gil's shirt and a belt belonging to one of his guides.

The marines took a group of Indians as captives and left the next day. They were followed downriver by a group of Indians presumed to be from one of the other villages. At one point, when the expedition party believed the Indians were about to attack, the marines began firing their weapons.

The marines and Gil escaped to the coast with six hostages, apparently a family of two adults and four children. Later, the hostages were brought back to the village with another expedition. But the village was now deserted—stripped of any sign of life.

There is still some doubt as to whether the tribe was in fact the Yuri Indians. The Yuris are thought to have disappeared about 50 years ago. They once numbered in the thousands, but Colombian slave traders who sold the Indians into forced labor at Colombian rubber plantations may have wiped out the entire population.

45. NORSÄLVEN LANDSLIDE
Sweden   **April 12**

A clay and sand landslide covered an area of about 350,000 square feet to a depth of up to 20 feet. The slide cut off a road and blocked the Nor River, causing river waves about 10 feet high.

46. NAPA FIREBALL
Napa, California   **April 15**

A very brilliant white fireball with a reddish tail was seen by residents of the Napa Valley, California. First spotted about 30 degrees above the horizon, the head and three-degree-long tail were watched by Napa residents for two to four seconds before their view of the brilliant object was obstructed by trees. Several observers claimed to have seen a bright flash of light when the fireball reached the horizon. No ionization trail remained after the meteor's disappearance, and no meteoritic material was found.

47. SAMAR SPONTANEOUS SOIL BURN
Philippine Islands   **April 20**

A substrata soil burn in western Samar, Philippines, turned the lush green vegetation of that area to black cinders. The burned vegetation included an estimated 30,000 coconut trees.

The remarkable incident stemmed from two factors: an unusually severe drought and lumpy black soil composed of hydrocarbon materials. The dried soil, in fact, had become so highly combustible it could be ignited with a torch.

Once the underground fire began, the ground was warm to the touch; and, for weeks, the fields of Samar smoldered as though a volcanic eruption was imminent. Huge coconut trees, their roots burned through, fell to the ground, and the fields became covered with a thin gray ash—the residue of the burned hydrocarbons in the soil.

The soil-burning phenomenon is not unusual in the Samar area. It usually occurs every year, but had previously been confined to small areas. In past years, the earth fires were extinguished without much trouble. But, because of the prolonged 1969 dry season, the underground fire could not be controlled, and Samar inhabitants simply had to wait until the earth burned itself out.

Samar spontaneous soil burn, Philippine Islands. *Photo courtesy of Pedro J. Almoradie, Jr., and Bernardo Tolentino, Philippine Commission of Volcanology.*

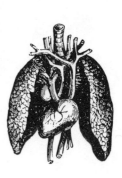

48. RINCÓN DE LA VIEJA VOLCANIC ACTIVITY
Costa Rica   **April 22**

Erupting an ash cloud up to 10,000 feet, the Costa Rican volcano, Rincón de la Vieja, became active April 22. Gas eruptions continued until April 29. While Rincón de la Vieja was ejecting gas and clouds of ash, simultaneous activity was reported at two other Costa Rican volcanoes, Poás and Mt. Arenal.

Another ash eruption occurred at Rincón de la Vieja on September 20.

Samar spontaneous soil burn, Philippine Islands. Acres of coconut palms and other fruit trees collapsed from the burning soil in western Samar.

49. SHANTUNG TIDAL WAVE
People's Republic of China   April 23

A gigantic tidal wave, generated by winds up to 200 miles per hour, flooded a 45-mile stretch of coastline in the heavily populated Shantung Province of the People's Republic of China. The tsunami was the greatest wave to strike China in 80 years.

The wave swept the Shantung coast in the area where the Yellow River empties into the Gulf of Pohai between China and Manchuria. The wave created a massive flood, with water levels reaching three feet in some inland areas as far as 13 miles from the coast. More than 1,100 square miles of farmland were flooded and thousands of buildings were destroyed in coastal areas. No estimate of deaths or injuries was released by the Chinese government, but some 100,000 people were left homeless.

The initial wave, reported as 22 feet high, destroyed the sea banks of four prefectures in the Shantung province. According to one report from Hong Kong: "The winds and waves smashed houses like matchsticks, damaged larger buildings, and destroyed some 1,000 square miles of rich farmland."

Subzero weather and a snowstorm hampered rescue operations by millions of Chinese soldiers and civilians.

50. BELFAST METEORITE FALL
Kilrea, Northern Ireland   April 25

Two meteorites were recovered about 30 miles apart after a brilliant and widely observed fireball passed over England, Wales, and northern Ireland.

The first fragment, weighing 0.4 kilogram, penetrated the asbestos roof of a constabulary shed in Sprucefield, a half-mile west of Lisburn, Northern Ireland. Some of the asbestos fused into the crust of the fallen fragments.

The other meteorite was found in a meadow in the townland of Bovedy, near Kilrea, county of Londonderry. The stone, weighing about 5 kilograms, was found in a small pit about 30 yards from a farmhouse. The meteorite penetrated more than a foot into the clay-rich soil, with an angle of penetration about 40 degrees from the horizontal.

The fireball created by the meteorite's passage through the atmosphere was first seen over southeast England and North Wales. It passed overhead at Cardiff and Aberystwyth at an estimated height of 40 miles. Continuing in a northwest direction, it then passed west of Anglesey at a height of about 25 miles, crossed the South Down coast of Northern Ireland midway between Ardglass and Newcastle, and finally passed overhead at Tyrella Beach on Dundrum Bay.

Nearly all observers reported a tail between five and ten degrees long, as well as the presence of a number of smaller bodies (usually three) following the main body.

The sonic boom and shockwave were heard and felt in many parts of Northern Ireland. In some places, the shockwave was so intense houses and cars were shaken. A girl in Bangor, who was tape-recording bird songs at the time the bolide passed overhead, recorded the subsequent boom. Also, observers in Dublin, Belfast, Antrim, Lisburn, Dungannon, and Kilrea reported a "swishing" or "rushing" sound heard as the fireball passed over.

Over Wales, the fireball appeared to be blue-green in color; it was "fiery white" over Northern Ireland with an intensity equal to, or brighter than, the full moon. Other fireball colors reported were red and orange.

Fragmentation of the fireball was seen by a number of people. One observer near Aberystwyth said the meteor broke into about six pieces, but all were extinguished in a fraction of a second. Just north of Dublin, an observer reported one part broke off and continued in flight with the main body. In County Down, observers said the fireball broke into three parts.

Near to the impact point of the large stone, observers reported three distinct fragments of similar size moving across the sky. Other observers witnessed a spray of bright red particles.

51. *HAMILTON TRADER* OIL SPILL
Liverpool Bay, England   April 30

A collision at sea involving the oil tanker *Hamilton Trader* was responsible for depositing 500 to 600 tons of oil into Liverpool Bay, England. More than 1,000 seabirds, including guillemots, razorbills, and red-throated divers, were casualties of the spill.

Twelve days after the oil gushed from the *Hamilton Trader*, it washed ashore along some 150 miles of the Cumberland coast. Areas of the Lancashire and southwest Scotland coastlines were marred with oil as well.

Norsälven Landslide, Nor River, Sweden. Aerial view upstream showing the blocked river and the landslide in the wood area. *Photo courtesy Curt Freden, Sveriges Geologiska Undersokning, Stockholm.*

52. SACSAHUAMÁN MINIATURE SCULPTURE FIND

Cuzco, Peru   **April**

A number of tiny sculptures, most no more than seven to nine millimeters long, were unearthed at the Sacsahuamán fortress in the Peruvian city of Cuzco. The objects were carved from slate and black andesite in the forms of animal heads, corncobs, fruits, and human faces. The facial features of the tiny sculptures indicate the pre-Columbian sculptors were of the Quechus or Aymara groups.

53. NEW JERSEY FISH–CRUSTACEAN MORTALITY

New Jersey   **March–May**

Scuba divers investigating the New Jersey coastal waters from Sea Bright south to Surf City discovered a veritable graveyard of crustaceans and fish. At depths of less than 100 feet, the divers reported widespread death of sea life. The sea life affected included cunner, ocean pout, black sea bass, tautog, rock crabs, lobsters, and mussels. In addition, great migrations of fish and crustaceans from the area had been reported by other divers.

The probable cause for the high sea life mortality and the migrations was the low level of dissolved oxygen in the water. However, the cause of the low level was not determined.

54. DAVAO LAND RISE

Philippine Islands   **May**

An earthmound on the base of a mountain ridge near the northern bank of the Mahanob River on the island of Davao in the Philippines suddenly rose approximately 15 feet in height. The land bulge was about 150 feet long and ended abruptly in a steep slope.

The phenomenon was first noticed in early May. The bulging at the ridge base rose rapidly, reaching a height of some 12 feet in less than three weeks. At about the same time, earthquakes shook the region and sounds similar to the crushing and sliding of rocks were reported.

A stream that once flowed by the foot of the ridge was uplifted by the land rise so that the riverbed was exposed. Trees remained standing on the newly formed hillock, but they now generally inclined eastward. Higher, the slope split in large cracks, some measuring about nine feet wide. The absence of any increase in ground temperature, vegetation dehydration, or sulphurous gases discounted the possibility that the event was volcanic in origin.

Most likely, the land rise was the result of "slumping," a kind of subsurface ground failure in which huge blocks of earth slide down a slope through an unusual combination of pressure at the top and reduced support at the bottom. Erosion, man-made excavations, and earthquakes may contribute to slumping. All these factors, in fact, may have contributed to the slumping and land rise on Davao.

The foot of the slope had been washed away by flooding of the stream a few months previously. Then, persistent rains created additional weight on the top of the slope. Both conditions made the area ripe for slumping. Finally the earthquakes also may have played a part in triggering the earth movement, even though some of the tremors passed before the actual movement began.

When the blocks slid along the slip surface, boulders were crushed against each other, accounting for the sounds heard by the people at the scene. As the material piled up at the foot of the slope, a natural earthmound was formed, forced up by internal pressure beneath the surface.

55. POÁS VOLCANIC ACTIVITY

Costa Rica   **May 3**

On May 3, Poás, a volcano often linked with activity at Mt. Arenal, erupted violently, sending vapor some 5,000 feet above the crater. Ten days after the vapor ejections, another significant eruption ejected vapor, steam, and black smoke. On the same day, an earthquake of Richter magnitude 6.5 shook the northwest provinces of the country.

On May 30, the activity at Poás intensified. Pulsating emissions of gases and pyroclastic materials were observed gushing from the crater. These emissions reached heights of 50 to 80 feet and occurred on the average of every six minutes, and were accompanied by large black clouds some 1,000 feet high.

Eruptions of ash and increased seismic activity continued for several months, but the volcano did little damage outside of the immediate crater area.

56. HOUSTON FIREBALL II

Houston, Texas   **May 3**

While flying at an altitude of 7,000 feet, the

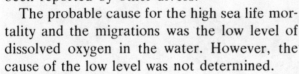

pilot of an Eastern Airlines Boeing 727 aircraft reported seeing a meteor about 30 nautical miles west of Houston, Tex. The pilot actually saw two objects, one following the other. The first was bright green, surrounded by a kind of halo; the second was white and much less brilliant. Both objects had very small tails.

The pilot's view of the objects was somewhat obscured by thin clouds that may explain the halo effect. The pilot reported that the objects had flat trajectories, nearly parallel with the horizon. Just before the objects burned out, they elevated slightly, as though they had "skipped."

The sighting lasted about 10 seconds. There was no explosion or increased brilliance at the end of the burn, and both lights went out simultaneously.

57. AMCHITKA EARTHQUAKE
Aleutian Islands, Alaska   May 14

An earthquake of magnitude 6.7 on the Richter scale rocked the Aleutian Islands south of the Alaskan mainland. The strength of the tremors gave officials reason to believe the quake would produce tsunamis, but no destructive tidal waves were reported in the days following the event.

58. SALO LANDSLIDE
Salo, Finland   May 19

A 15-foot-long section of ridge equaling some 2,000 cubic meters of earth tore loose from the bank of the Salo River near Salo and slid for a distance of 150 feet. A few years previously, the land on the opposite side of the river had sunk about 30 feet.

59. SANDY HOOK FISH KILL
New Jersey   May

Thousands of dead and badly chilled blowfish were found on the beaches along the New Jersey coast from Sandy Hook to Manasquan Inlet. The cause of death was believed to be a radical drop in water temperature—from 55 to 40 degrees Fahrenheit in a few minutes—at the shallow depths that the fish had recently occupied.

60. PISCATAQUA RIVER OIL SPILL
Portsmouth, New Hampshire   May

As it moved out to sea, a barge struck a bridge abutment on the Piscataqua River in New

Hampshire, spilling more than 200,000 gallons of oil into Great Bay and Rye Harbor. Oil flowing into the intertidal zone killed both young and adult clams, intertidal oysters, clam-worms and estuarine shrimp. The shrimp and clam-worms are a primary food source for fish in the region. The ecological balance was affected in another way: the rockweed that holds small organisms which are at the beginning of the food chain died because of the oil.

61. JUNIN LAGOON MARINE POLLUTION
Junin Lagoon, Peru   May

The marine life in the Lagoon of Junin in Peru was reported disappearing because of mine washings deposited into the waters. The decomposition products of the mine washings—sulphur, acids, and toxic foams—caused the death of one-third of the Junin frog, bird, and fish populations.

62. UTHAITHANI SITE DISCOVERY
Thailand   May

Three sets of prehistoric colored wall paintings and a 3,000-year-old bronze axe were found in four caves in the Uthaithani province of Thailand. Two of the caves contained landscape paintings, while the other two contained pictures of men and fish drawn in curious "X-ray" style.

63. SOUTH CHINA FLOODING
People's Republic of China   May

Summer rains raised the flood level of the Yangtze and Pearl rivers in Anhwei Province in southern China to the highest level in history. The floods were extensive throughout the land surrounding the rivers, and more than 200,000 soldiers were mobilized to fight the flood and give aid to the homeless.

64. SOOTY TERN HATCH FAILURE
Dry Tortugas, Florida   May–June

The failure to hatch of an estimated 98 percent of the eggs in a colony of sooty terns, *Sterna fuscata,* was reported this season on Bush Key, one of the islands in the Dry Tortugas group off the southern tip of Florida.

A colony of noddy terns, *Anous stolidus,* on the key also did not hatch normally. The cause of the sooty tern hatch failure is unknown, although one scientist claimed that sonic booms might have played some role in the unprecedented occurrence.

Later in the year, a banding expedition went to the Tortugas. While some 7,000 to 25,000 sooty terns are usually banded, the scientists this year were able to band only 300.

65. MASSACHUSETTS BIRD KILL
Southern Massachusetts   **June–July**

Large numbers of dead birds were found on the south shore of Massachusetts Bay during the week after the area had been sprayed from a helicopter with Baytex, an organophosphate pesticide.

Eight birds were analyzed. A University of Massachusetts research team analyzed the tissue of eight birds known not to have died from natural causes. The tissue showed the presence of lethal chlorinated-hydrocarbon pesticide residues, primarily DDT. From the specimens tested, it was concluded that the bird mortality was caused by residues of DDT, Baytex, or a combination of both.

66. NEWPORT CLAM IRRUPTION
Newport, Rhode Island   **June**

Thousands of small surf clams, *Spisula solidissima,* were washed ashore at Easton Beach in Rhode Island. A subsequent investigation found no cause for the irruption. A similar influx of clams occurred in the Newport area during the summer of 1909, causing investigators to believe the 1969 beaching was "a sporadic and random phenomenon."

67. BECEJ NATURAL GAS ERUPTION
Yugoslavia   **November 1968–June 1969**

While drilling near the Yugoslav town of Becej on the Hungarian border, a natural gas explosion lifted the drilling rig off the ground. Immediately following the explosion and for the next six months, a powerful stream of air roared through the drilling hole.

The strange geyser produced a loud, eerie whistling noise that did not stop until April 9, 1969. Unfortunately, the silence—otherwise a relief for people who had heard the unearthly whistling for nearly six months—turned out to be an ominous sign. The next day, large amounts of natural gas—93 percent carbon dioxide and 7 percent methane—began seeping from the well. The gas poisoned five people before the area was evacuated. Some relatively unsuc-

cessful attempts were made to make the gas less toxic by concentrating it with water and to disperse it with turbofans.

Then, on June 4, the gas seepage stopped suddenly. A scant half-hour later, the natural gas volcano erupted again; this time creating a huge mushroom cloud and tossing a chunk of earth the size of a house 300 feet into the air. The original drilling hole became a crater nearly 500 feet in diameter.

Becej natural gas eruption, Becej, Yugoslavia. *Photo taken on June 4th. 1969. by the Foreign Photo Service. Hungarian Telegraphic Office. Budapest.*

The explosion apparently stopped any further gas eruptions. But low rumblings and murmurings could be heard from the crater; and, at one point, mud was ejected to a height of over 60 feet. All activity stopped abruptly on June 7.

68. UBINAS VOLCANIC ACTIVITY
Moquegua, Peru   **June 1**

Beginning on June 1 and continuing into July, Ubinas Volcano began erupting, with an ejection of sulphurous gases and dense smoke. Some ejected ash also caused damage to crops in nearby farmland.

69. NORTH ISLAND FIREBALL
North Island, New Zealand   **June 3**

A number of North Islanders witnessed a spectacular fireball as it moved across the sky from south to north. Information from three reports suggested that the meteor's end-point was about six miles seaward of Fanal Island. Loud explosions were heard about three minutes after the meteor could no longer be seen, and calculations derived from these and other reports suggested that the fireball exploded below the ozone belt, about 34 miles above the earth. No meteoritic material was found.

70. MISSOURI MASTODON FIND
Missouri   **June 4**

The bones of a mastodon, rarely uncovered in the midwestern United States, were found at a construction site in northwest St. Louis County, Mo. The bones were estimated to be 10,000 to 30,000 years old.

The bones were uncovered by the driver of a bulldozer cleaning the site of a future office complex. The bones were 23 feet below the grade. A research team was called in and they excavated the bones—several vertebrae, some ribs, long bones, parts of tusks, scattered bone fragments, and parts of teeth. The scientists identified the bones as the prehistoric remains of a mastodon from the teeth.

In addition to the bones, traces of charcoal and two snail shells were found at the same level as the mastodon remains. The field investigation was sponsored by the St. Louis Museum of Science and Natural History, where the bones are now on exhibit.

71. IOWA FIREBALL
Iowa   **June 5**

Persons in Illinois and Iowa watched a spectacular fireball travel east to west across the two states. Although widely observed over an area of 350 miles, no sonic phenomena were reported.

One observer in Grant, Iowa, said the meteor separated into three equally bright objects, with one seemingly moving faster than the other two. The rear object disappeared a few seconds after the split, while the second fragment vanished immediately afterward. The remaining fragment faded and slowly disappeared from sight about 30 degrees above the horizon. The color of the object was white or blue-white and passed across the sky in ten seconds.

72. WEYMOUTH OIL SPILL
Weymouth, Massachusetts   **June 7**

A barge transporting heating fuel spilled an estimated 100,000 gallons of its cargo into the Fore River Channel, thus producing a three-mile-long oil slick off the Massachusetts coast. The oil floated out of the Fore River Channel into Quincy Bay and onto the Weymouth shoreline. Crabs, clams, and other mollusks were reported to have been killed by the oil pollution.

73. ARARAQUARA FIREBALL
Araraquara, Brazil   **June 11**

A number of people from the Araraquara District of Brazil saw a fireball cross the sky in a northwest direction, "giving the impression that it was about to fall somewhere along the horizon."

The report from Brazil called the meteor a "strange ball of fire" that followed a descending path until it disappeared. The object reportedly gave off intense light and had a bright red nucleus, something like incandescent metal. Beams of light of a bluish hue were also seen around the edges of the fireball.

74. MEDITERRANEAN SEA EARTHQUAKE
Eastern Mediterranean   **June 12**

An earthquake, with its epicenter located in the eastern Mediterranean Sea, was recorded on June 12. The quake reached magnitude 6.3 on the Richter scale, Luckily, no major surface waves were generated by the quake and destruction was slight.

75. ALMA OIL SPILL
Alma, Wisconsin   **June 16**

The barge *Jim Hougland*, carrying 567,000 gallons of Number Two home-heating fuel, sprang a leak after running aground in the Mississippi River near Alma, Wis. The Army Corps of Engineers estimated that some 40,000 gallons of oil spilled from the barge. Other estimates, however, varied from 54,000 to 400,000 gallons.

The oil covered approximately 80 percent of the water's surface for a distance of 20 miles along the river and caused extensive shoreline pollution and fish and insect mortalities. The Wisconsin Department of Conservation began an investigation of the immediate oil damage to wildlife.

76. RHINE RIVER FISH KILL
Rhine River, West Germany and the Netherlands   **June 19**

Approximately 40 million fish were killed by a highly toxic substance called Endosulfan somehow introduced into the Rhine River. The chemical is a chlorinated cyclic hydrocarbon marketed by the Hoechst chemical firm under the trade name of Thiodan. The source of the

Endosulfan, presumably introduced into the Rhine at a point near Bingen, West Germany, has not been traced.

Dead fish began to be noticed in the river at Bingen on June 19. A wave of dying fish continued to sweep down the Rhine toward the Netherlands, and Dutch authorities were duly warned. On the next day, Dutch officials took samples of the water and identified the alien toxic substance as Endosulfan. The samples contained from 0.44 to 0.70 microgram of the poison per liter of river water.

On July 7, the Endosulfan levels were considerably reduced and it was announced that the contamination was no longer dangerous.

77. VICTORIAN TREMORS
Australia   **June 20**

An earthquake shook Victoria, Australia, with a magnitude of 6.0 on the Richter scale. A number of aftershocks were recorded up to four days after the main tremor.

78. RUAPEHU VOLCANIC ERUPTION
North Island, New Zealand   **June 22**

On June 22, Mt. Ruapehu, in the center of New Zealand's North Island, erupted violently. The explosive eruption from the crater was of ash, not lava. Several small avalanches of hot ash melted snow, creating some mudflows. Ash deposits were observed up to 10 miles from the crater, and enough water was expelled during the eruption to lower the level of the crater lake 10 to 20 feet. Boulders up to 10 feet in diameter were scattered up to 500 yards from the crater's rim.

According to the official report filed by an aerial inspection team, "An eruptive ash cloud had been emitted and was directed towards the northwest by a strong wind blowing at the time. Ash fell along a strip extending down the northwest side of the volcano and for some miles away from the foot.

"It is suspected that the ash on the northwest side which covered the Whakapapa glacier was also hot and melted further snow, sending mud-flows down the glacier and for some distance down the Whakapapanui and Whakapapaiti valleys," said one report.

Other reports described the sonic phenomena accompanying the volcano. One New Zealander felt a shockwave—actually a compressioned air wave—at his home some five miles from the Ruapehu crater. Two or three rolls of a low, long, rumbling sound were heard by one person. At approximately the same time, still another observer is reputed to have seen flashes of light and a shower of sparks and heard loud rumbles from the direction of the crater. This observer described the sight as a "massive fireworks display."

Later in the day, a hail of ash and scoria fell on a skiing area nearly three miles away from the crater. Finer ash deposits were subsequently observed up to nine miles to the northwest of Ruapehu.

The mudflows caused some damage to structures near the mountain. One flow partly demolished a shop and cafeteria on the ski grounds. A shelter hut immediately above the crater was flattened, presumably by the eruption blast. A lahar, generated by hot water from the lake or by hot ash flow, flowed to the Wha-

Ruapehu volcanic eruption, North Island, New Zealand. View showing thick ash deposit and course of mudflow. *Photo courtesy Dr. J. Healy, New Zealand Geological Survey, Rotorua.*

kapapaiti and Whakapapanui rivers. In the Whakapapaiti, nine miles from the crater, three to four feet of sand and scoria were deposited following the eruption, much of the material probably being from the stream's valley. In the Whakapapaiti the water level rose five to six feet above normal during the lahar passage. The two rivers join and feed the Wanganui River where a kill of trout, eels, and insects was reported. The ecological effects of the volcano were many. The rivers in the area were made poisonous by volcanic ash, sulphurous materials, and acids. Dead trout and eels, some weighing up to five pounds, were found along the river banks. Some were still alive, but their skins were peeled, suggesting a considerable quantity of acid in the water.

No sign of insect life was found in the water touched by volcanic emissions. Because insects are the food source for the trout, it appeared that a re-establishment of a trout population in the affected streams would be very difficult. The fish mortality continued well beyond the immediate area of the volcano. One observer 32 miles away from the volcano counted two dead fish every minute floating downstream.

An eruption of Mt. Ruapehu in 1945 caused the same kind of ecological damage in the North Island rivers and river valleys.

79. COOK INLET OIL SPILL II
Cook Inlet, Alaska   **June 23**

Following a devastating oil spill in March, considerable leakage of fuel oil from a Liberian tanker left a wake of contaminated water the full length of Cook Inlet. After investigation, however, there did not seem to be any significant danger to marine life from the oil.

80. CORUSQUEIRA LANDSLIDE
Azores   **June 25**

Several tons of rock and soil slid down a 180-foot escarpment at Corusqueira, in the Azores, and destroyed several houses before rolling into the sea. On July 27, another landslide occurred at the same place, but caused no damage. The Azores are islands in the North Atlantic off the coast of Portugal.

81. INDIANA FIREBALL
North-Central U.S.A.   **June 26**

Shortly before midnight on June 26, a fireball was seen over Indiana and Michigan. One

observer in East Lansing, Mich., reported the object crossed the face of the moon. He said the fireball was green in color and as bright as the moon it crossed. The fireball's green color, the observer also noted, provided a dramatic contrast to the moon's pale whiteness.

Before it disappeared, the fireball reportedly turned reddish pink with a "glowing ember in its center." The "ember" continued to glow after the fireball seemed to go out, and then broke into six to eight fragments, each glowing faintly for a while. There was a momentary trail left behind the fragments, but no vapor trail was reported.

82. CHILI RIVER CONTAMINATION
Chili River, Peru   **June**

The Chili River in Peru was contaminated from continuous disposal of acid residues and wastes from tanneries along the banks between the towns of Grau and Bolivar. Survival of certain species of aquatic life was reported threatened by the contamination. The toxic substances in the river had already caused the disappearance of the small frogs and fish, according to reports.

83. CASTROVIRREYNA SINKING
Ticrapo, Peru   **June**

According to reports received in Lima, the village of Ticrapo, in the district of Castrovirreyna, Peru, was slowly sinking back into the earth. The phenomenon was caused by substrata faulting that created great cracks in the earth along a circle with a two-mile circumference. Adobe houses were said to be slowly inclining and the imminent collapse and destruction of the buildings was reported.

84. RALEIGH BAY BIRD KILL
Cape Hatteras, North Carolina   **June**

Although no accurate figure is available, some estimated thousands of shearwaters, long-winged sea birds, died from unknown causes off the coast of North Carolina. Particularly affected was the species *Puffinus gravis,* known as the "greater shearwater," which died in about 50 times the number of the *Puffinus griseus,* or sooty shearwater.

Dying shearwaters were first noticed from Beaufort to Cape Hatteras in early June. By the middle of the month, reports were received from Wilmington to Nags Head and from as far north as Assateague Island, Maryland. Tissue from sooty shearwaters was analyzed for chlorinated hydrocarbon and organophosphate pesticide residue, but the levels found were below those considered lethal.

Another series of tests were made on the birds, but the cause of the mortality remained a mystery.

Raleigh Bay bird kill, Cape Hatteras, North Carolina. *Photo courtesy of The Raleigh Times.*

85. WATERFOWL KILL—BOTULISM
U.S.A.   July–September

Outbreaks of botulism resulted in a major waterfowl mortality across the United States. In July, substantial numbers of waterfowl died in Pennsylvania and Maryland. And, in Michigan, botulism resulted in the deaths of more than 1,000 ring-billed gulls and a number of common terns. In August and September, major waterfowl deaths were reported in South Dakota and California. At the same time, a less serious mortality was reported in North Dakota.

The various types of botulism are caused by different strains of the bacterium *Clostridium botulinum*. The bacteria, which are anaerobic and therefore do not live in the presence of air, thrive in warm stagnant waters. Insects and fish, the major food source of the birds, contain the botulinus toxin produced by the bacteria. The toxin results in the paralysis and ultimate death of the birds.

86. TOLFA TREMORS
Italy   July 2

A series of tremors were recorded with their epicenter in the region of Tolfa, Italy. The Rome Institute of Geophysics registered the first shock at 6.0 on the Mercalli Scale, the others at 5.0 and below. Three aftershocks were recorded, bringing the total number of recorded seisms to 12.

87. LEINE RIVER POLLUTION
Niedersachsen, West Germany   July 3

Large quantities of fish were found dead in the Leine River between the towns of Gottingen and Alfeld, West Germany. The fish kill was caused by the high quantities of oil, grease, and organic solvents phenol and benzol found in the river.

88. WINDWARD PASSAGE FLOATING ISLAND
Windward Passage, Caribbean Sea   July 4

The U.S. Navy destroyer escort *John D. Pearce* reported an island afloat in the Windward Passage between Cuba and Haiti some 60 miles south of the Guantanamo Naval Base in Cuba. The island, reported moving through the strait at about 2.5 knots, was 15 yards in diameter and covered with some 10 to 15 palm trees, each about 30 or 40 feet tall.

One observer said the island "looked as though it were held together by a mangrove-type matter." The report said there might have been earth on the island, but it appeared to be mainly composed of dead foliage, bush, and grass. The roots of the tall trees were not visible. No animal life was observed on the island.

The report of the island—first considered a navigational hazard—was an event of major ecological significance. Ecologists regarded the island as a possible clue to the transmigration of some species of plants and animals.

On July 18, a team of Smithsonian Institution ecologists prepared for a helicopter flight to the island. But four days earlier, a report from the Guantanamo Naval Base stated that the island had broken up. What remained was a single 40-foot tree floating upright with about ten feet of its length above water.

On July 19, the U.S. Coast Guard made an intensive six-hour search for the island, but nothing was sighted. The island was presumed sunk, and the ecological expedition was terminated.

89. TUNIS FIREBALL
Tunis, Tunisia   July 6

The strange blue disc the size of the moon seen moving north across the Tunisian sky was first believed to be a fireball.

The object was reported to have changed shape, from circular to elliptical, to have disappeared, reappeared again, and then finally to have exploded, "lighting the sky and giving all the characteristics of a nuclear explosion."

Later reports, however, confirmed that the disc was actually a cloud formed by burning oxygen escaping from a rocket used in atmospheric testing.

The rocket was launched from the island of Sardinia in the Mediterranean and reportedly reached an altitude of 150 miles. The oxygen cloud formation stretched over an area of 18 miles.

90. UNITED KINGDOM BIRD KILL
Southeastern England   July 6–9

Over 1,500 birds of various species died during a period of sudden and heavy rainstorms over southeastern England. Examination showed no evidence of either disease or poisoning. However, scientists found that the dead birds seemed, for some reason, to have been weakened, or "out of condition." Researchers felt the bad weather may have caused the death of the birds, but "some other cause" was responsible for the weakening of the birds, since a mortality of this size was without precedent.

91. YELLOW SEA EARTHQUAKE
Tientsin, China   July 18

An earthquake in the Yellow Sea near Tientsin, People's Republic of China, was reported at a 7.7 magnitude on the Richter scale. No tsunami was generated by the quake.

92. CHARLES RIVER FISH KILL
Watertown, Massachusetts   July 19

Approximately 10,000 fish died from a highly toxic, oil-based chemical spilled into the Charles River upstream of the Watertown dam near Boston. The substance, believed to be a plasticizer, came from Mutrie Motor Transportation, Inc., Waltham, Mass.

An investigation into the fish kill was sponsored by the Massachusetts Department of Natural Resources.

93. PERUVIAN TREMORS
Huancayo, Peru   July 24

Twenty villages were cut off by floods, scores of houses were destroyed, two bridges collapsed and water and electricity services were interrupted following a series of earth tremors in central Peru. A report from the Reuters news agency said 17 tremors of varying strengths had been registered in the region around the city of Huancayo, about 200 miles south of Lima.

Huancayo's 200,000 people were left without drinking water, and the city's main power plant was damaged by the tremors, leaving some areas without electricity. The city, rocked by the heaviest seismic activity since 1947, was declared to be in a state of emergency.

Two snowpeaks in the Huaytapallana glacier area collapsed due to the quakes and crashed into Lake Lasintay and the Shullcas River, raising the water level by 15 feet and flooding the surrounding countryside. About 20 villages between Huancayo and Chili Fruta were cut off by the floodwaters. Two bridges were swept away and a large area of cultivable land was ruined.

94. GYPSY MOTH INFESTATION
Northeastern United States   July–August

Gypsy moth larvae, *Porthetria dispar,* defoliated approximately 150,000 acreas of forest in Connecticut, New Jersey, and New York. Both chemical and biological methods were employed in efforts to stop the larvae advance and to eliminate those in the infested areas. A continuing investigation was begun by the Plant Pest Control Division of the U.S. Department of Agriculture.

Growing concern over the use of pesticides and the side effects of DDT on animal life, however, caused officials to seek other means for controlling the pests. The gypsy moth, for example, has at least seven natural enemies. One small wasp from India attacks the moth eggs, other insects from Spain eat the larvae stages, while the Carabid beetle eats both larvae and eggs. A spray has also been developed that interferes with the caterpillar's ability to eat, and a special virus that attacks only the gypsy moth is being tested.

Another more complicated way of eradicating the moths is the "sterile release method." In this approach, massive numbers of male moths are sterilized through gamma-ray radiation and then released, thus reducing the population of the next gypsy moth generation.

95. SOUTHERN LAKE MICHIGAN ALEWIFE MORTALITY
Illinois, Indiana, Michigan   July–August

The 1969 alewife mortality in southern Lake Michigan is believed to have been caused by a rapid change in shoreline water temperature. Water temperatures measured varied from 65 to 75 degrees Fahrenheit at the surface and from 46 to 58 degrees Fahrenheit at the bottom. Freshwater alewives are thought to be intolerant of sharp temperature changes of this kind. All sizes of adults were found in the mortality, with yearling alewives accounting for about 9 percent of the total.

96. MOLUCCA PASSAGE EARTHQUAKE
Indonesia   **August 5**

An earthquake recorded at magnitude 7.0 on the Richter scale struck a remote part of Indonesia. There were no reported tsunamis generated by the quake.

97. OHIO–INDIANA FIREBALL
Ohio and Indiana   **August 5**

A meteor widely observed over the midwestern United States was seen traveling in an east-west direction, while changing color from white to orange. The meteor produced a glowing white cloud visible for approximately two minutes. No sonic phenomena were reported. The fireball was seen in Ohio, Indiana, West Virginia, Kentucky, and Michigan.

98. LAS CRUCES FIREBALL
Las Cruces, New Mexico   **August 9**

The final display of an incredibly brilliant fireball was observed from a point near the New Mexico State University golf course in Las Cruces. The object, much brighter than a full moon, lit up the night sky as it traveled from southeast to northwest. The fireball was seen for less than three seconds. No sound phenomena were reported.

99. SOUTH AFRICAN OILING OF PENGUINS
Cape Peninsula   **August 10**

Over 30 penguins, *Sphenicus demersus,* came ashore at Cape Peninsula covered with oil. Since there was no record of an oil spill in the area, it was concluded that the penguins were contaminated by "chronic" oil, that is, a persistent and permanent oil residue in the open sea. An investigation was sponsored by the South African National Foundation for the Conservation of Coastal Birds.

100. KURILE ISLANDS EARTHQUAKE
Kurile Islands, U.S.S.R.   **August 1**

An earthquake of Richter magnitude 7.8 struck the Soviet Kurile Islands, a chain extending from northern Japan to the southern tip of the Kamchatka peninsula. About a minute before the main shock, a foreshock occurred at magnitude 6.0. Within the following 24 hours 30 aftershocks were recorded. Tsunamis were also reported, but no serious damage was caused from either the earthquakes or the tsunamis.

101. PANAMA MOTH MIGRATION
Panama   **August 1**

Migrations of the moth *Urania fulgens* are not a surprising event in Panama. The migrations normally occur every year, moving generally westward in the dry season and eastward in the wet season. But the August migration was unusual for its size. A marked increase in the number of moths sent literally millions of the moths over the small Central American country. The increase in the moth population was presumably the result of a very successful breeding season.

102. HUNGARIAN MUDFLOW
Tatabanya, Hungary   **August 15**

Due to abnormally heavy rains, a highway leading to Budapest near the town of Tatabanya was covered by a layer of mud nearly 2,000 feet long and nearly a yard deep. Another mudflow covered a road near Vertes at the same time.

103. *GIRONDE* OIL SPILL
English Channel   **August 19**

In the English Channel off the coast of Brittany, the French oil tanker *Gironde* and the Israeli cargo ship *Harbashan* collided, destroying two of the *Gironde's* oil tanks and emptying 1,500 tons of fuel oil into the sea. An oil slick four miles wide and seven miles long resulted from the crash.

By August 25, strong winds had moved the slick onto the beach between the French coastal villages of Erquy and Val-André, contaminating an area some three miles long and 1,000 feet wide. Powdered chalk and sawdust were used in an attempt to clean up the oil.

Although some reports claimed no biological effect, there were other accounts to the contrary. One report claimed that scores of cormorants and other sea birds were found on the beach and treated for oil damage. A colony of some 3,000 to 5,000 birds were said to be in the Bay of St. Brieuc near the collision point. Other parts of the bay were known to be very rich in shellfish. However, no reports indicated whether the oil damaged the bird colony or the shellfish.

104. PRINCE GEORGE METEORITE FALL
British Columbia, Canada   August 21

Visual sightings and seismic records indicated that a large meteorite had fallen near the city of Prince George late in the evening of August 20. Seismographs at four different stations recorded the impact, and residents in the Prince George area felt strong local tremors, but no meteoritic material was found.

The fireball traveled from north-northwest to southeast, as seen from Prince George. Sonic effects were noticed also, but residents said they seemed mild compared to the earth tremors. The suspected impact area is mostly hilly, containing a large mature spruce forest.

105. HUEHUETENANGO LANDSLIDE
Guatemala   August 28

Eighty persons were believed killed when a landslide of mud and stones damaged the Guatemalan villages of San Antonio and Las Nubes. The slide was caused by heavy rainstorms.

106. GALAPAGOS ISLANDS TREMORS
Galapagos Islands   August 30

A shallow earthquake of magnitude 6.0–7.0 on the Richter scale recorded on August 30 may have been related to the Fernandina caldera collapse of 1968. A second quake followed on the same day, accompanied by a series of aftershocks.

107. HIGH AIR-POLLUTION POTENTIAL
Eastern U.S.A.   August

The National Air Pollution Control Administration Division of Meteorology reported extremely high air pollution in the eastern section of the United States during August. The condition was considered unusual because of its long duration and large area coverage.

108. EUROPEAN BIRD MIGRATION SCARCITY
Northern Europe   June–August

Substantial population decreases of two spe-

cies of birds that normally summer in Great Britain and Northern Europe were reported in 1969. Reduced numbers of the sand martin, *Riparia riparia,* were seen in Belgium and Britain; and populations of the white-throated warbler, *Sylvia communis,* were considerably lowered in Britain, France, and Scandinavia.

Reports from Great Britain showed that the white-throat warbler counts were dramatically less than any year since 1962. A Smithsonian Institution project determined that a relatively high rate of viral infection had struck the bird population during its 1968 autumn passage in Egypt.

109. GUYANA ARCHAELOGICAL FIND
Guyana, South America   June–August

While digging in the Amatok region on the lower Cayuni River, employees of the Guyana Timbers company found a perfectly oval stone bowl, several stone implements, and stone dishes. They were found below other potsherds buried 9 to 12 inches below the surface. Also found were several rock paintings.

Imessi-Ile boulder slide, Nigeria. A portion of the boulder's 80-meter-long slide path. *Photograph by Sylvester Kindzeka.*

110. IMESSI-ILE BOULDER SLIDE
Nigeria   September 1

A huge rock of granite, weighing 200 to 300 tons, slid from the side of an inselberg near Imessi-Ile in western Nigeria. The boulder, 15 feet high and 25 feet in diameter, cut a path 250 feet long. Vegetation and soil were pushed in front and to the side of the boulder. The sound of its passage was heard three miles away and caused widespread alarm that the mountain might be collapsing. The boulder slide occurred during a heavy rainstorm.

111. MEIJE ROCKFALL
Southeastern France   **September 2**

A rockfall from the southwest side of the 12,000-foot Mt. Meije covered an area of almost two square miles. The rocks fell an estimated 4,200 feet. The mountain is located approximately 30 miles east southeast of Grenoble, France.

112. AMCHITKA PASS EARTHQUAKE
Aleutian Islands, Alaska   September 12

An earthquake northeast of Amchitka Island was recorded at a magnitude of 6.6 on the Richter scale. An hour and 40 minutes before the main quake, four foreshocks were recorded at magnitudes of 5.2, 5.4, 5.4, and 5.3. An aftershock of magnitude 5.7 occurred on the same day.

113. ALEUTIAN ISLANDS VOLCANIC ERUPTION
Aleutian Islands, Alaska   September 12

On the same day as the Amchitka Pass earthquake, Kiska Volcano, on Kiska Island in the Aleutians, began erupting. The volcano was visible from Amchitka, 50 miles away, and observers saw volcanic ash shoot 1,200 feet in the air and clouds of steam rise to about 12,000 feet. They also saw the volcano spout flames and spit out lava.

During the days after the initial eruption, however, observation of the volcano from Amchitka—and from aircraft—was hampered because of poor weather. Activity apparently increased on September 16. Photographs showed steaming blocky lava, but it could not be discerned whether it represented a new flow.

114. WEST FALMOUTH OIL SPILL
Massachusetts   September 16

When the barge *Florida* ran aground off Chappaquoit Point in Buzzards Bay, an estimated 175,000 gallons of diesel fuel spilled into the sea. Along the five-mile stretch of shoreline between Nyes Neck and Chappaquoit Point, 24 species of fish and shellfish died in large numbers as a result of the spill.

Lobsters, shrimp, crabs, shellfish, and seaworms were affected by the spill, but hardest hit were the scallops, which died by the thousands even in areas where there was no evidence of oil. Seagulls and shorebirds were not directly harmed by the spill, but the loss of some food sources would eventually seriously affect them.

Shellfish beds in the area of contamination were officially closed by the Massachusetts Department of Public Health. An investigation of the accident was undertaken by the Division of Marine Fisheries, Massachusetts Department of Natural Resources.

115. POLIČKA METEORITE FALL
Czechoslovakia   September 16

A Czech woman watched a bright fireball flash across the sky. Suddenly, she heard a sound similar to an airplane motor, followed by a loud crash. A meteorite—one of the year's most significant finds—had crashed into her farmhouse only 140 feet from which she stood.

The meteorite smashed six roof tiles and destroyed a roof support beam before breaking into two pieces. One piece weighing 60.1 grams fell into the loft; the other, a 755.2-gram specimen, rolled down the roof into the front yard. The larger piece measured 13.5 by 7.0 by 5.5 centimeters.

Although 815.3 grams of the stone were recovered, scientists estimated the entire fallen mass equaled about 840 grams. Samples of the meteorite, a chondrite, were distributed to laboratories all over the world. A remaining 686.1-gram piece was placed on display at the National Museum in Prague.

116. MIAMI SNAIL INFESTATION
Miami, Florida   September

An unsuspecting tourist introduced giant African snails, *Achatina fulica,* to the Miami area several years ago, thus inadvertently causing a crisis that still exists.

In 1969, the snails proliferated over a 13-block residential area of about 400 acres in northwest Miami, devouring any form of green vegetation they could find.

The snails are hermaphroditic and have no natural enemies in Florida, thus they can reproduce at an alarming rate. For example, the lifespan of a giant snail is approximately five years, and each snail can produce 400 to 600 offspring a year. Worse yet, the snails often go into a kind of hibernation where they remain hidden in a cool, dark spot for more than six months at a time.

When the snails awake from their suspended animation, or estivation state, they are vora-

Miami snail infestation, Florida, U.S.A.

hibernate in the chrysalis stage to escape inclement weather. But, during the weeks before the swarm, frequent, short, heavy thunderstorms broken by periods of bright sunshine apparently confused the insects and a simultaneous hatching of extensive numbers of the butterflies took place, resulting in the profusion of butterflies.

118. MURCHISON METEORITE FALL
Victoria, Australia   September 28

A daylight fireball, seen and heard by thousands of Australians, produced one of the year's most scientifically interesting meteorite showers. About 30 pieces of meteoritic material, weighing a total of ten pounds, were recovered in and around the town of Murchison. Most were found lying on or near a dirt road, but one—a 1.5-pound specimen—ripped through the roof of a barn, narrowly missing two men at work stacking hay.

The men who recovered the stone from the haystack reported that it was cold and smooth, unusually heavy for its size, and covered with a black crust as though scorched in a fire. They also said the stone gave off a peculiar odor similar to denatured alcohol, which later proved to be pyradine.

The widely observed fireball was seen to break into three pieces before disappearing into puffs of smoke; sonic phenomena were also heard. The stones and fragments of the meteor were discovered along a mile-wide track extending five miles southeast of Murchison.

The meteoritic materials were parts of a single carbonaceous chondrite, fairly compact and at least somewhat heterogeneous. Since the stone fragmented in air, however, the fractured surfaces had incipient fusion crusts on them. The carbonaceous chondrite is a rare type of meteorite, rich in carbon materials. More than a year after the stones showered the Murchison countryside, scientists at the National Aeronautics and Space Administration announced they had detected the presence of amino acids, the building blocks of protein, in the meteoritic material. This was the first time the amino acids, a basic ingredient of living matter, had been found in meteorites.

cious eaters. To halt the plague of snails, the Florida Division of Plant Industry baited the area with the molluscide metaldehyde-tricalcium-arsenate. This bait produced a 95 percent kill within 72 hours. In one week, 2,500 dead snails were found; two weeks later, another 5,000 were picked up. By the end of the year, about 17,000 giant snails had been killed.

Although the giant snail population had been diminished, it was not entirely eradicated. By late 1970, Florida officials reported finding five to ten snails a week. Because of the snail's unusual powers of reproduction and hibernation, another outbreak or infestation of snails remains an annual threat.

117. TRINIDAD BUTTERFLY INVASION
Trinidad, West Indies   September

Unusually large swarms of butterflies filled the Trinidad skies in September. The butterflies, *Urania leilus* and *Phoebis statira,* usually

119. GREAT BRITAIN BIRD MORTALITY
Western Great Britain   September

Over 6,000 dead and dying birds washed

ashore on the western coast of Scotland during the early days of September. Many of the birds were oiled and the first assumption of naturalists was that the birds were victims of an oil spill.

However, it soon became clear that a major bird mortality was taking place on the coasts of England, Ireland, and Wales and that oil was only one of a combination of factors responsible for the deaths. High level traces of toxic chemicals, such as polychlorinated biphenyl, were found in the birds.

The majority of the some 10,000 birds that died in the three-month period were guillemots, *Uria aalge*. Other species affected included cormorants, shags, puffins, razorbills, and gannets.

120. HUANCAYO EARTHQUAKE
Huancayo, Peru   October 1

Many villages and towns in the area of Huancayo City were seriously damaged by a major earthquake and the 1,037 recorded aftershocks that followed it. Many homes were destroyed and landslides blocked all major roads in the area. Some parts of highways disappeared completely.

Although superficial cracks were seen on the ground surface, great fissures were reported along both sides of Mt. Huaytapallana. The fissures extended for several miles and created landslides that heavily damaged villages below.

121. SANTA ROSA EARTHQUAKE
Santa Rosa, California   October 2

An earthquake of Richter magnitude 5.6 struck Santa Rosa, California, creating five aftershocks: three on October 2, and two on October 6. A long fissure was created by the main tremor, and then was increased to 600 feet long by the October 6 aftershocks.

122. CAPE PROVINCE EARTHQUAKE
Capetown, South Africa   October 6

The villages of Wolseley, Ceres, and Tulbagh, lying in a small triangle in the mountains 50 miles northeast of Capetown, South Africa, were the hardest hit by the most severe earthquake to strike that region since 1932.

The quake jarred the coastline from Capetown to Durban, collapsing buildings and killing at least ten people. Reports from Wolseley stated that hardly a house was left standing there, while several people were killed in Tulbagh, the victims of collapsing houses.

The quake was enormous, even though it lasted about 15 seconds. Officials at the South African Magnetic Observatory said the tremor was so powerful that it knocked the seismograph needles "off their pivots." One Capetown apartment dweller said the noise of the quake was "like an express train roaring through the place."

Fires broke out in the villages after the first severe tremors raged in the mountains. The earthquake was so extensive that tremors as far north as Uppington, 500 miles from Capetown, were reported to have damaged buildings there.

123. ST. LOUIS SPIDER INVASION
St. Louis, Missouri   October 8

Masses of strange, sticky, threadlike material floated across St. Louis the afternoon of October 8, causing general consternation. The bubblelike material, ranging from "dime size to ten-foot globules," drifted into the heart of the city, sticking to grass, metal, and cement.

The white substance was first thought to be a synthetic precipitate from heavy air pollution, but it was later identified to be nothing more than spider webbing.

The webbing came from an unusually large number of "balloon spiders." The females of this species crawl to the tops of trees and weave webs into which they discharge their eggs. The webbing is then severed and dispersed by the wind. Apparently the weather conditions on the afternoon of October 8 were such that enormous numbers of webs wafted across St. Louis. At the height of the invasion, the air over the city was filled with the webs at a density of almost one per square foot.

124. OZARKS FIREBALL
Midwest U.S.A.   October 9

Very early on the morning of October 9 an exceptionally bright meteor over Oklahoma, Kansas, and Missouri was photographed by nine of the 16 camera stations of the Smithsonian Astrophysical Observatory's Prairie Network. The network is operated primarily to determine accurate geographic locations of possible meteorite falls.

The Ozarks Fireball was so brilliant, however, that one photograph taken at Pleasanton, Kan., some 35 miles from the fall, showed some image reversal. The reversal suggested that the magnitude of the fireball was -20.

Ozarks Fireball, southwest U.S.A.

ly constructed highway between Westport and Perth, Canada. On October 13, one man identified 36 garter snakes, 16 black rat snakes, three water snakes, and one ribbon snake killed; other carcasses were either too mangled or deteriorated to be identified.

Several causes for the snake mortality were suggested. Recent road construction may have disturbed the snakes' natural habitat, the warm road surface may have attracted large numbers of the reptiles, or, for several reasons, a periodic migration of the snakes may have been undertaken.

126. CANLAON VOLCANIC ERUPTION
Negros Island, Philippines   October 11

The Canlaon Volcano eruption on the evening of October 11 was characterized by a mild ejection of ashes and white smoke. An earthquake was felt on the island at the same time. Two days later, however, the volcano sent clouds of ash 15,000 feet into the air during a half-hour eruption accompanied by cannonlike explosions. Heavy rains following the eruption turned the accumulated volcanic ash on the ground into a mudflow that blocked roads and caused some damage on the island.

127. CALIFORNIA CHANNEL EARTHQUAKES
California   October 22

A small swarm of earthquakes from the San Andreas fault area hit the southern California coast during the fourth week of October. The first quake (Richter magnitude 5.5) struck off the west coast of Santa Barbara County on October 22, while the second quake (magnitude 5–5.5) shook Laguna Beach on October 24.

The quake apparently created a fissure on a seaside cliff at San Pedro more than two feet wide and 100 feet deep. The fissure cut two homes in half and knocked two others off their foundations. The fissure was probably caused by the tremors, but a high watertable resulting from heavy rains the winter before had eroded and weakened the cliff structurally.

128. GOLDEN-CROWNED KINGLET MORTALITY
Lake Huron, Ontario, Canada   October 23

Thousands of dead golden-crowned kinglets, *Regulus satrapa satrapa,* were found along five miles of Lake Huron beachfront near Kettle

Calculations based on the photographic data showed that before the meteoroid entered the earth's atmosphere, it was moving around the sun in a strongly elliptical orbit near the orbit of Venus, but still part of the asteroid belt. The object's heliocentric motion was from west to east like a planet, and its orbit was inclined only 12.6 degrees to the ecliptic.

Initially the meteoroid must have had a mass of several tons. It penetrated the atmosphere to within 15 miles of the ground and caused sonic booms over a wide area. But the complete disappearance of the object following the explosions was considered virtual proof that no sizable fragment reached the earth. The material was therefore assumed to be very fragile and perhaps cometary in origin.

125. WESTPORT SNAKE MORTALITY
Westport, Ontario, Canada   October

Between 100 and 200 snakes were killed by motor vehicles on an eight-mile stretch of new-

Point, Ontario. While on their southerly migration October 22 and 23, the birds apparently were caught in a snowstorm. Within a few days the bird carcasses disappeared into Lake Huron as a result of continued stormy weather in the area.

129. BANJA LUKA EARTHQUAKE
Yugoslavia   October 26

A severe earthquake felt throughout Yugoslavia and southern Austria caused one death and an estimated 100 injuries. The quake also caused a great amount of property damage to buildings near its epicenter at Banja Luka, Yugoslavia. The shock was severe enough to be felt as far away as Sicily.

Several aftershocks of considerably less force were recorded within a day's time. By October 28, an additional 14 tremors had struck the Banja Luka area, making a total of 27. There were also three main aftershocks following the swarm, the first at magnitude 4.0–5.0 Richter on November 3, the second at 4.0 on November 4, and a third at magnitude 4.8 on December 31.

130. KOVACHI SUBMARINE VOLCANO
Solomon Islands, South Pacific   October

An underwater volcano known as Kovachi in the Solomon Islands boiled the sea, sent clouds of steam and fountains of water up to 500 feet into the air, and coated the sea with floating lava. The activity began on October 28 and did not stop until well into 1970.

On the first day, eruptions shot columns of water 100 feet into the air every 30 seconds. During the next two days, these ocean geysers were reduced in size by almost half and then finally stopped, leaving only a boiling, churning sea as evidence of the volcano. At the same time, however, a scum of red-brown lava covered an area of 70 nautical miles.

Then, on November 3, great water spouts again shot up to 500 feet at 30-second intervals. For the next two days, great water eruptions up to 200 feet high occurred at least once a minute. On November 10, only one eruption was reported, but it was 300 feet high and produced a huge steam cloud, as well. As of late December, activity still continued, with water spout

Kovachi submarine volcano. Solomon Islands. South Pacific. *Photo courtesy of Dr. R.B.M. Thompson. Geological Survey. Homiara. British Solomon Islands.*

ising to 100 feet every five to ten minutes.

In early December 1970, the volcano was observed to be welling up very gently; and, by December 23, the activity had renewed and water spouts 100 feet high rose every five to ten minutes.

131. MOUNT TAAL VOLCANIC ERUPTION
Philippine Islands   October–December

On the morning of October 29, Mt. Taal began a two-month period of extensive volcanic activity. The initial eruptions, coming at 90-minute intervals, were of ash and small fragments. By October 30, these ejections reached heights of 600 to 700 feet. The erupting vent widened to 300 feet in diameter, with ash and lava fragments ejected 1,500 feet into the air.

By November 10, the volcano's physical environment had changed due to the massive flow. The volcano built a new cone southwest of Volcano Island on the slope of a 600-foot-high crater created the previous year. The conelet was reported to be about 200 feet high. The cone's lava front advanced even farther, into the area where a lake had collapsed in the 1966 eruptions. The Taal eruption is an unusual case for volcanologists, for it is only the second volcano in recorded history which emitted more lava than ashes.

The volcano continued to pour out steam and ash and to open up new vents throughout the fall. By December 9, the volcano's activity was reportedly limited to the ejection of incandescent materials and ash to heights of about 1,500 feet.

On December 29, 1970, the Philippine Volcanology Commission reported that the situation remained "critically serious" at Mt. Taal. The report listed profuse steaming, a fresh landslide, an active rift passing through the main crater, and an earthquake.

132. BEZYMIANNY VOLCANIC ERUPTION
Kamchatka, U.S.S.R. October–November

Bezymianny Volcano in the heart of the Kamchatka peninsula has been in a state of eruption since March 1965. From 1966 to 1969, three new spines were formed on the west flank of the dome; and, during October and November of this year, seismic activity increased radically and included a total of 130 earthquakes.

From October 11 to October 25 an eruption occurred, characterized by an eruption cloud reaching a height of a little over a mile and by small *nuées ardentes* avalanches down the east flank of the dome for a distance of over a mile. In November, the volcanic activity was limited to normal gas emission from the dome.

133. SOUTHEAST IRAN EARTHQUAKE
Iran   November 7

A major earthquake of magnitude 7.0 on the Richter scale struck southeastern Iran on November 7. Only mild tremors, however, were felt in the area's main cities, and no damage resulted.

134. MENTAWAI TRENCH EARTHQUAKE
Sumatra, Indonesia   November 21

A powerful earthquake of Richter magnitude 7.7 reportedly shook the Atjeh Province, Sumatra, and was felt as far away as Malaysia. The epicenter of the quake was located off the west coast of Sumatra.

135. NEW JERSEY–NEW YORK CROSSBILL IRRUPTION
New Jersey–New York   November

Migrations of white-winged and red crossbills, which normally winter no farther south than southern Canada and northern New York, came as far south as Long Island and northern New Jersey. About 400 of each crossbill species, *Loxia leucoptera* and *Loxia curvirostra,* had been reported in the New Jersey area alone. Scientists attributed the southern irruption to a scarcity of pine and spruce seeds in the north.

136. OSTROV KARAGINSKIY EARTHQUAKE
Kamchatka, U.S.S.R.   November 22

A strong earthquake, with its epicenter located near the island of Ostrov Karaginskiy off the coast of the Kamchatka peninsula, was recorded at 7.5 magnitude on the Richter scale. The quake caused minor tsunamis up to three feet high along the shores of the Aleutian Islands.

137. MOUNT LOKON VOLCANIC ERUPTION
Celebes, Indonesia   November 27

Mount Lokon erupted when a sudden explo-

sion sent a thick, dark cloud 1,500 feet into the air and ejected incandescent rocks that subsequently fell to earth near the crater. Three earthquakes were recorded during the initial eruption.

On the next day, an explosion ejected pyroclastics for 15 minutes and produced a cloud 1,200 feet high. Sulphurous mud and sand fell on nearby villages, withering vegetation and killing fish in some ponds.

On both December 1 and 9 there were several explosions, usually accompanied by dark eruption clouds and light ash falls. Until the middle of the month, small-scale eruptions occurred at 20-minute intervals and were accompanied by light ash falls. During the last half of December, however, the eruptions became more intense. Violent explosions on December 23 and 25 resulted in mud rains and ash falls.

138. GREAT BRITAIN SEAL MORTALITY
Cornwall, Great Britain   November

The deaths of at least 40 young seals from three colonies in Cornwall were reported in November and December. The colonies, each consisting of 100 to 200 adult seals, normally maintain a birthrate of about 100 to 150 pups a year so that the dead pups represented a 25 percent mortality rate.

Some scientists claimed such a rate is not unusual; nonetheless, no satisfactory explanation of the deaths could be given.

139. VICTORIA MOUSE PLAGUE
Australia   December

Millions of mice overran areas of Victoria and New South Wales, damaging crops and causing a general alarm throughout Australia. The increase in mouse population started in early 1969, but it was not until December that farmers became aware that the mouse population had reached plague proportions.

In one part of Victoria the mice numbered in the millions, with an estimated mouse density of more than 200 per acre. Farmers trapped as many as 300 to 400 in a single night; and, in some places, the common rodents were described as "a moving carpet." The Malee region attacked by the mice covers 11 million acres, representing one-fifth of the area of the state of Victoria.

One report from the town of Cuyen, 200 miles northwest of Melbourne, said the town "was

alive with mice." In other areas, the number of the animals killed on the roads became so extensive that speed limits had to be reduced to prevent the chance of cars skidding on the carcasses.

Australia had suffered mouse plagues in 1917, 1932, and 1949. Then, as in 1969, the main damage caused by the mice was to crops. Rough estimates made indicated that the mice might have eaten up about 20 percent of the year's rice, corn, and wheat. The plague was the first in which the mice actually attacked growing rice and seed crops.

Widespread use of poisons was discouraged, for it proved ineffective against the hordes of mice and possibly more dangerous to livestock. Among the poisons used were strychnine, phosgene, organophosphates, and parathion. None of the poisons seemed to be completely successful against the mice; but, when bulkhead storages were enclosed with a fence and then fumigated, the mice were controlled and their numbers greatly reduced. Where polythene was used in storage areas, little damage was done by the mice.

The cause of the 1969–1970 mouse plague was probably unseasonal rains, which provided an abundant food supply and produced an increase in the mouse population. The numbers of mice that lived through the winter thus increased; and, with large supply of food in the spring, the population grew even larger, until it reached plague proportions.

Some communities did not report a large mouse population. Apparently the rodents did not migrate, but rather contained their population explosions to those areas where a large food supply was available.

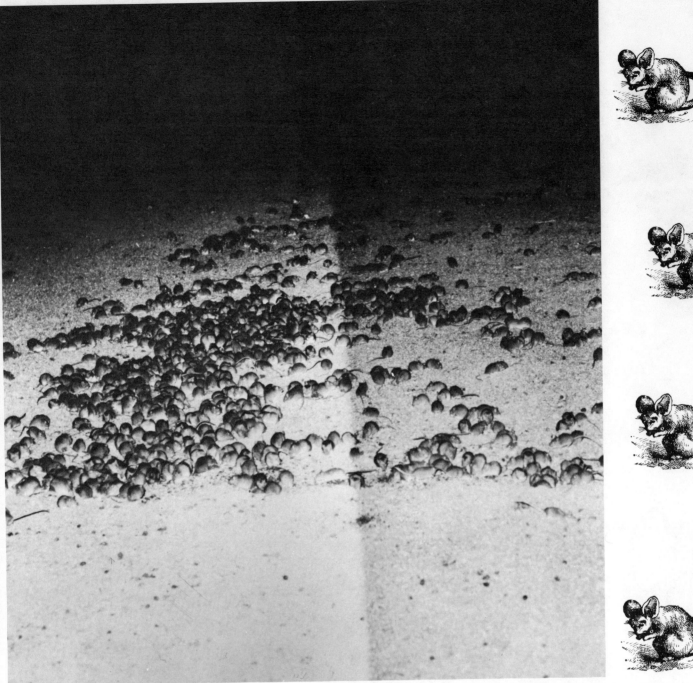

Victoria mouse plague, Australia. Thousands of mice invading grain storage bins in Australia. *Photo courtesy Keith Turnbull Research Station, Victoria, Australia.*

140. HAWAIIAN ISLANDS STORM SURGE

Hawaii   **December 1**

An unusually high surf—very possibly the highest in several decades—struck the northern and western coasts of the Hawaiian Islands on December 1. The surge was generated by a great low pressure area in the North Pacific. The storm, originally centered about 1,800 miles north of Hawaii, created monumental waves as it moved slowly eastward. The waves rolling into the Hawaiian beaches were reported from 30 to 40 feet high. Tern Island in the French Frigate Shoals was reported to be under four to six feet of water as a result of the storm.

141. LEEWARD ISLANDS EARTHQUAKE

Guadeloupe, Caribbean Sea December 25

The largest earthquake ever recorded in the eastern Caribbean Sea registered magnitude 7.8 on the Richter scale at stations on the Leeward Islands. An aftershock of Richter magnitude 6.5 occurred shortly after. There were no reports of tsunamis, property damage, or injuries connected with the quake.

142. SPANISH MOSS MORTALITY
Southeastern U.S.A.   January–December

An unexplained mortality of Spanish moss, *Dendropogon usneoides,* occurred in Florida and certain areas of the southeastern United States during the year. Agencies in ten counties, from the northeastern to the western part of Florida, first reported the incidences of dying moss. Tests made by state officials ruled out insects or plant disease as the cause of the mortality. Although a definite cause was not established, some officials believed that man-made air pollution may have been responsible.

143. INDO-PACIFIC STARFISH PLAGUE
Indian and Pacific Oceans   1960s

During the decade of the 1960s, increasing numbers of Crown of Thorns starfish destroyed extensive areas of coral reefs in the Indian and Pacific oceans. The starfish destroyed the reefs by eating the outer living layers of coral which formed protective living shields, thus exposing the reefs to the continuous erosive action of the sea.

The staggering increase in the Crown of Thorns population and the steady weakening of coral reefs from Guam to Australia to islands in the Indian Ocean greatly alarmed scientists, who feared the destruction of coral reefs could alter both geography and the ecology of the Indo-Pacific region. As one marine biologist put it: "If the starfish explosion continues unchecked, the result could be a disaster unparalleled in the history of mankind."

Indo-Pacific starfish plague. Indian and Pacific oceans.

The ugly, 16-armed starfish first appeared in unusually large numbers in 1963 at Green Island, a resort area on the edge of the Great Barrier Reef in Australia. Within less than a decade, the starfish were reported throughout the Indo-Pacific area, ranging as far east as Hawaii and as far west as Ceylon. This species of starfish may grow two feet wide. And one adult starfish can devour an area of coral its own size in less than 24 hours. Once a coral reef is attacked by the starfish, it crumbles under the ocean's pressure and is washed away. The consequences of widespread coral reef destruction are devastating, for it sets off a frightening chain of events. Without the coral reef's protection, the sea life that abounds there will disappear, followed by the island life (human and animal) that depends on the reef for food supply. Eventually, as the reef disappears and the island is exposed to open sea, the island itself could be washed away.

The damage done in the 1960s by the Crown of Thorns was extensive. For example, on Guam, where the starfish population increased from almost negligible numbers to more than 20,000 in only three years, nearly 24 miles of coral reef was destroyed.

Major investigations by a variety of regional and international groups attempted to evaluate the causes and extent of damage. The Crown of Thorns has two natural enemies—the giant triton shell, and, ironically, the coral itself, which feeds on starfish larvae. Some scientists believed that if enough coral was destroyed—perhaps by underwater blasting or atomic tests—it might have become capable of coping with the increasing numbers of starfish, thus becoming the predator instead of the prey.

The giant triton shell, the greatest enemy of the starfish, is also a favorite souvenir for tourists. The numbers of the large tritons could have been depleted by collectors to the point where the starfish could breed unchecked. Other factors besides the exuberance of shell collectors—including blasting, dredging, and chemical pollution—might also have been responsible for the reduction in the triton population.

Editor's Note: By the early 1970s, marine biologists had revised their rather pessimistic predictions for the future. A decrease in starfish populations was noted in various observation sites throughout the oceanic area. The decrease apparently was unrelated to any control efforts; rather, it seemed a natural die-off.

Apparently the spontaneous and still inexplicable increase in starfish population was a cyclic phenomenon that might not be repeated for many decades. For the present, at least, the reefs seem safe.

144. SEABIRD EGG CONTAMINATION
Southern California   Late 1960s

High concentrations of pollutants such as DDE, an inert residue of DDT, appeared to be causing abnormally thin-shelled eggs among seabirds along the California coast. The eggs are so delicate they often break before they ever hatch. Brown pelicans, common murres, egrets, cormorants, and other seabirds all seemed to be affected by the pollutants.

145. JAPANESE BAMBOO KILL
Japan   1969–1970

In Japan, *ma-dake*, or *Phyllostachys bambusoides,* is the principal species of cultivated bamboo. The *ma-dake* blooming cycle averages between 60 and 120 years, depending on the temperature of the area where it grows. In 1960, bamboo plantations in Japan began to flower for the first time in over 100 years.

The flowering, however, is no reason for joy in Japan. Indeed, the flowering is fatal to the plants. About a year after the flowers appear, the greenish-yellow stalks wither and die. Another 10 to 15 years are needed before the stalks can again reach a harvestable stage.

By 1966, 30 percent of Japan's total *ma-dake* crop had bloomed and died. By the end of 1969, almost 50 percent of the bamboo crop, or about 100,000 acres of bamboo forests, was affected. The last major flowering of such scope occurred in 1846.

146. SCOTLAND NECK BLACKBIRD IRRUPTION
North Carolina   1960–1970

For five years, a remarkable roost of blackbirds and starlings plagued the town of Scotland Neck, North Carolina. The birds, estimated to number 3 million in 1969, created serious social, economic, and health problems for the town.

First, the birds created an incredibly loud and constant noise at sunrise before they departed for the fields where they fed on peanut and grain crops. More dangerous were the health problems caused by bird droppings. The

Japanese Bamboo Kill, Honshu-Kyushu, Japan. *Photo courtesy Dr. F.A. McClure, Smithsonian Institution.*

Japanese Bamboo Kill, Honshu-Kyushu, Japan. *Photo courtesy Dr. F.A. McClure, Smithsonian Institution.*

droppings, high in nitrogen, so fouled the town's canal that health officials compared the contaminated water to raw sewage. When the droppings dried, the powder was carried by the wind, creating the potential for an outbreak of histoplasmosis, a viral lung infection, in humans. Bird lice on dead birds created another health hazard.

In March, the town began a harassment campaign designed to drive the birds to another roost. The birds were sprayed with detergents from airplanes: the detergent, Tiergital 15-5-9, removes the natural oils from the birds' wings and prevents them from flying, leaving them exposed to the elements. Recorded starling distress calls were used, as were explosive devices, in an effort to scare the birds away. The project, which cost several thousand dollars, seemed somewhat successful. But, at the next roosting season in the fall, shocked townspeople discovered that the blackbird and starling populations had doubled rather than decreased. Some estimates went as high as 12 million birds. This time the townspeople tried to smoke the birds out of their roosts at night, but this and other methods used were unsuccessful. Perhaps the only sure way is to chop down the trees, and force the birds to find a new place to roost.

There are approximately 200 major blackbird and starling roosts (a major roost contains one million birds) in the United States, with most located in the southeast. Most roosts are in unpopulated areas, but occasionally they are found near communities where they create major nuisance and health problems. The total blackbird and starling population in the United States is thought to be about 500 million birds. Pest birds, including the starling and blackbird, are responsible for an estimated $50 to $100 million in damage to crops and vegetation in the United States each year.

147. SWEDISH SNOW POLLUTION
Southwest Sweden   December–January

On Christmas morning in 1969, families in the Lake Vattern area in southwestern Sweden awoke to find the snow had turned black. That night more snow, followed for two days by freezing rain, caused the snow cover to turn a grayish color, giving the landscape a dull and ominous appearance.

The substance discoloring the snow was gritty, oily, and difficult to remove from clothing. Animals wandering in the forest left white tracks in the gray-black snow.

The area was immediately investigated by the Stockholm Ecological Center, and tests showed that the black substance had an alarmingly high level of DDT and other toxic industrial pollutants. The snow apparently had cleared the air of these pollutants. Unfortunately for Lake Vattern residents, the cleansing snowfall brought the ugly chemical ooze back to earth.

148. BRITISH ISLES SEABIRD MORTALITY
Great Britain   December–February

The deaths of approximately 15,000 sea birds in northeast Britain resulted from oil pollution. The mortality was noticed in mid-December 1969 and was generally concentrated in southeast Scotland around the mouths of the Firths of Tay and Forth.

Over 30 species of birds were affected by the oil which was brought to the British shore by prolonged east winds in the North Sea. The great majority of birds killed were guillemots (Uria aalge), although other species affected included razorbills (Alac torda), little auks (Alle alle), puffins (Fratercula artica), eiders (Somateria mollissima), and common and velvet scoters (Oidemia nigra and fusca).

The mortality ranked after the Torrey Canyon disaster and Irish Sea kill in the fall of 1968 as the largest yet recorded in the British Isles. The source of the oil was thought to be either the residue of fuel-tank cleaning on merchant vessels or the drilling operations for oil and gas in the central North Sea in which oil is used as a lubricant for the drills and also in hydraulic jacks.

Most of the British beaches were only slightly polluted by the oil.

EVENT REPORTS   1970

Martha's Vineyard bird mortality, Massachusetts, U.S.A. *Photo courtesy of the Record American. Boston. Mass.*

1. BRAZILIAN COFFEE RUST
Brazil   **1970**

A young Brazilian scientist discovered evidence of coffee rust, *Hemileia vastatrix,* in January. Throughout the year, the rust scattered over an area of some 200,000 square miles, including the states of Bahia, Minas Gerais, and Espirito Santo.

Some trees were thought to have had the disease for several years. Quarantine and eradication were considered impossible due to the extent to which the rust had spread. Although there were no estimates of crop loss, the outbreak threatened the entire coffee industry in Brazil where more than two billion coffee trees are under cultivation. The last appearance of coffee rust in the Western Hemisphere was in Puerto Rico in 1963, when it was eradicated in its early stages.

2. TONGA SUBMARINE VOLCANIC ACTIVITY
South Pacific   **January 3**

Submarine volcanic activity was reported off Falcon Island in the South Pacific as evidenced by a discoloration of seawater in a 500-acre area, but no new island was formed as a result.

3. TUNISIAN NEMATODE INFESTATION
Tunisia   **1969–1970**

Heterodera rostochiensis, commonly known as the "golden eelworm," is the most serious nematode pest in potato farming. In many countries, including the United States, there are very rigid quarantine regulations against the pest. Yet, somehow, the eelworm had been introduced into Tunisia during 1969. Because eelworm cysts can remain dormant in the soil for many years, the pest is "still in a phase of build-up in Tunisia." Potato farming prospects for the future, thus, do not look bright. Losses in potato and tomato cultivation are expected.

4. KUN-MING EARTHQUAKE
Peoples' Republic of China   **January 4**

A severe earthquake, estimated at between 7.8 and 8.0 magnitude on the Richter scale, reportedly shook Yunnan Province between the city of Kun-ming and the Burmese border. Although no definite information on the effects of the quake was known, the general Yunnan-Burma border is not considered to be densely populated. Still, Kun-ming is a city of approximately one million people.

A radio report on January 8, which said Chinese leaders sent a "cable of sympathy" to the people of the affected area, indicated that the earthquake probably was responsible for the deaths of a number of people and for extensive property damage as well.

5. LOST CITY METEORITE FALL
Oklahoma   **January 4**

A 22-pound meteorite found outside the small farming community of Lost City, Okla., 45 miles east of Tulsa, turned out to be an object of historic significance. It was the first meteorite ever to be recovered in a search guided by trajectory information computed from photographic data, and it was the second meteorite whose orbit around the sun, prior to entering the earth's atmosphere, was determined from photographic observations.

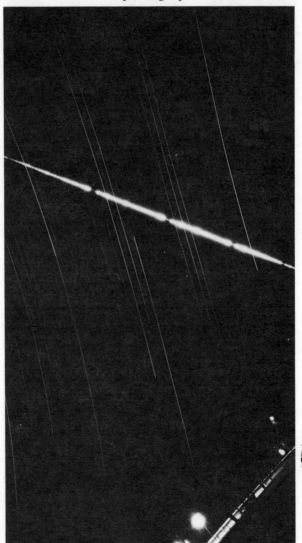

Lost City meteorite fall, Oklahoma, U.S.A.

The find was made by staff members of the Smithsonian's Prairie Network, a system of 16 automatic cameras in seven states that was set up to photograph fireballs and aid in the recovery of meteorites. The Lost City Meteorite, discovered in a snow-covered field five days after the fireball was photographed over northeastern Oklahoma on the night of January 4, was the Network's first successful recovery of freshly fallen material.

The fireball, brighter than a full moon, was seen from as far away as central Nebraska. Traveling in an east-southeast direction, it caused sonic booms heard from Tulsa to Tahlequah—60 miles apart. Network films of the meteorite's descent were quickly analyzed; and, with the help of a computer to compare photos from two different stations, the impact point was predicted to be a spot three miles east of Lost City. On January 9, the 22-pound specimen was found only a half-mile from the predicted impact point.

The specimen was flown to Washington, and in the Smithsonian's laboratories was identified as a bronzite chondrite. On January 17, a second 10-ounce fragment, half buried in the sod, was found by a local farmer while searching for a lost calf. Two more fragments were later found, thus making the total mass recovered 17.3 kilograms. Also, within 18 hours of the fireball sighting, a plane flew at 60,000 feet from Oklahoma downwind to Atlanta, Ga., and back, collecting upper-air samples in a search for meteoritic particles. The attempt proved successful and a large amount of ablation particles was collected.

6. PACAYA VOLCANIC ERUPTION
Guatemala   **January 7**

After months of moderate activity, the Pacaya Volcano in Guatemala erupted, sending long tongues of fire high into the night sky. The magnitude of the activity had been increasing for several months. For example, on December 15, a regular eruption produced ejecta, bombs, and scoria.

7. MOUNT ULAWUN ERUPTION
New Britain, New Guinea   **January 9**

The eruption of Mt. Ulawun (meaning "the father") devasted surrounding forest areas with its *nuée ardente*, or glowing cloud, emissions. The eruption was the most powerful recorded including even severe activity in 1915.

The volcanic activity occurred in three general phases: an initial vapor emission phase, followed by an explosive phase, and finally a quiet effusive and vapor emission phases. Ulawun, nearly 7,000 feet high, was a nearly perfect cone bare of vegetation above the 4,000-foot level. Before the January explosions, the central crater was 1,200 to 1,300 feet in diameter and sections of its irregular floor were over 300 feet deep.

The eruption produced marked changes in the summit's topography. The strong explosive activity from the northern vents of the volcano caused some cone growth and in-filling of the original crater.

8. DAVAO EARTHQUAKE
Philippine Islands   **January 10**

An earthquake caused a power blackout throughout the city of Davao on the Philippine island of Mindanao, although there were no other reports of casualties or property damage. The quake, which lasted 25 seconds, was recorded at Richter magnitude 7.5. The quake's epicenter was located 60 miles southwest of Davao.

9. FLORIDA BEACHED WHALES
Florida   **January 11**

For reasons not fully understood, approximately 150 false killer whales, *Psuedorca crassindens,* swam onto beaches along a two-mile stretch at Fort Pierce, Fla. Although similar whale mortalities are common throughout the world, they are unusual on the Florida coast. The most recent whale stranding occurred on one of the Florida keys in 1966 when 60 pilot whales came ashore.

The false killer whales ranged between 15 and 20 feet long, with an average weight of some 1,500 pounds. Efforts by the Florida Department of Natural Resources Marine Patrol failed to pull the whales back into deeper water. As soon as the whales were free, they immediately turned around and swam back onto the shore faster than the patrol boats could head them off.

Blood samples were drawn from six of the stranded animals before they were towed out to sea. The results of test showed no sign of diseases. Although indications of kidney and liver dysfunctions were found, this could have been due to the prolonged period the whales spent out of the water—more than 24 hours.

A second attempt at towing the beached whales was more successful. Eventually, some 50 whales were removed from the beaches and remained at sea. The other whales, however, died on the beaches from dehydration.

Whales usually travel in "herds," led by an old bull. The apparent urge of the Florida herd to reach the land baffled scientists. According to one theory, the beaching may have been due to unusually cold weather. With the water temperature as low as 26 degrees Fahrenheit, the bull leader may have stranded the herd in his search for warmer water.

Another theory suggested that the herd's leader may have been diseased and headed for shore to die, only to be followed by the other whales.

No matter what reason brought the whales to shore, the result was tragic. Whales become disoriented when they are in shallow

house in Ucera, about 330 miles from Caracas. A farmer saw the fireball as it streaked across the sky, then heard an explosion when it struck the ground 240 feet from the farmhouse.

The farmer ran into the field and recovered a small stone—later identified as a 4.95-kilogram olivine-bronzite chondrite—that was still warm to the touch. The brownish object was 32 centimeters long and 15 centimeters wide. No impact crater was found.

11. KERMADEC ISLANDS EARTHQUAKE
South Pacific   **January 20**

An earthquake of Richter magnitude 7.4 shook the Kermadecs and other South Pacific islands approximately 1,100 miles north of Wellington, New Zealand. The quake was preceded by a 6.6 Richter magnitude tremor on January 8 in the same general area.

Florida beached whale mortality, Florida, U.S.A. *Photograph courtesy of Wometco Miami Seaquarium, Miami, Fla.*

water, or in sloping shorelines such as that at Fort Pierce. Under these conditions, the sonarlike natural navigation instincts of the whales became confused. Herd instinct then takes over, and the whales succumb to a form of mass hysteria, similar to a cattle stampede.

10. VENEZUELAN METEORITE FALL
Venezuela   **January 16**

A five-kilogram meteorite fell near a farm-

12. MYOJINSHO VOLCANIC EXPLOSION
Japan   **January–June**

A number of explosions at sea on January 29 producing water spouts 300 feet high signaled the eruption of the submarine volcano Myojinsho in the southern Izu Islands, Japan. The explosions were accompanied by a yellow smoke and a 900-foot-wide yellow belt of water containing suspended sulphur moved along the south side of the volcano.

Another larger eruption in the Myojinsho Reef occurred on February 7. Crew members of a fishing vessel witnessed great dark-red clouds of spray and seawater around the reef.

The submarine volcano erupted again on April 23, sending a water spout some 1,000 feet in the air, as well as producing another yellow sulphur belt and a floating pumice band 300 feet wide and 3,000 feet long.

13. *ARROW* OIL SPILL
Canada **February 4**

The oil tanker *Arrow,* carrying 3.8 million gallons of fuel oil, ran aground in Chedabucto Bay, Nova Scotia, causing the ship's forward tanks to rupture. Sixty-mile-an-hour winds carried the oil, escaping freely from the tanker, across the bay toward the shore, causing a three-mile stretch to become polluted.

The shifting winds scattered the oil slick in different directions in the bay area. Residents reported seeing numbers of birds, mostly merganser, scoter, and eider ducks, covered with the oil. In the intertidal zone, much destruction of marine life was evident, with crabs and limpets most seriously injured. Most other marine life, however, including phytoplankton, did not seem affected by the oil, which is relatively nontoxic compared to crude oil or kerosene.

By February 14, about 1.5 million gallons of oil had spilled from the disabled tanker. After a survey of the oil-contaminated coastline, the Canadian Wildlife Service estimated that 2,300 birds were killed by the oil. The survey became more difficult when snow covered the beaches and ice gathered in inlets and coves. Still, small numbers of oiled birds continued to struggle ashore and dead birds washed up. Oiled seals were also reported, and several persons had spotted young gray seals in woods about a half-mile from the shore. The seals possibly had become disoriented due to oil accumulations around the eyes and nostrils.

14. MARTHA'S VINEYARD BIRD KILL
Massachusetts **February**

About 1,000 birds were the victims of oil pollution possibly leaking from a sunken vessel. The birds were found dead on the shores of Martha's Vineyard, but there was no sign of oil on the beaches.

However, residual oil was found in the ponds beyond the beach areas, which led to the theory that the oil may have come from a sunken vessel or from an accident involving a tanker in November 1969. Most of the dead birds were loons, common eiders, and white-winged scoters. Birds found alive were cleaned with detergent and kept at a wildlife station on the island.

15. CHESAPEAKE BAY DUCK KILL
Maryland and Virginia **February–March**

An outbreak of cholera among waterfowl in the Chesapeake Bay area was first reported on February 16 when dead old-squaws were found on the shores of James Island, Md. By the end of March, dead ducks were found on beaches from as far south as Norfolk, Va., and as far north as Annapolis, Md.

Personnel of the Maryland Department of Game and Inland Fish and the Bureau of Sport Fisheries patrolled 76.8 miles of beach on both sides of the bay and collected a total of 4,780 dead birds of which 3,371 were old-squaws and 895 were white-winged scoters. The total losses were estimated at 20,000, and perhaps as many as 40,000, birds.

16. KODIAK ISLANDS OIL POLLUTION
Alaska **February–March**

At least 1,000 miles of coastline along the eastern side of the Kodiak Islands, as well as other Alaskan shorelines, were contaminated by an oil slick.

Approximately 500 marine animals, mostly fur seals, were believed to have been affected by the oil. Although no seal and sea lion mortalities were reported in the immediate vicinity of the oil contamination, the fur of seals in other areas showed oil stains. In some cases the oil contamination could not be cleaned by detergents or gasoline, thus creating an economic hardship for those hunters depending on seal hides for income.

Although the exact source of the oil damage was unknown, investigators suspect it may have been deliberate discharges of "slop oil" from tankships, possibly within the restricted 50-mile oil-discharge limit.

The total effect of the oil on the area's ecosystem is unknown. Of some 27 beaches surveyed in the area, 85 percent were either contaminated or contained oiled bird carcasses and remains. Besides sea mammals, virtually all species of crabs, shrimp, and bottom fish were exposed to the oil. The estimate of birds killed ran from 10,000 to 100,000.

17. *DELIAN APOLLON* OIL SPILL
Florida   **February 13**

An estimated 10,000 gallons of oil spilled into Tampa Bay when the oil tanker *Delian Apollon* ran aground. The oil rushed out of a three-foot gash in the tanker's hull and contaminated approximately 10 to 15 miles of shoreline in the St. Petersburg area.

Although most of the pollution was contained within the bay, oil did spread into the Gulf of Mexico. In the immediate area of the tanker, limited use of a detergent-type dispersant was employed to prevent further spreading of the leakage.

Delian Apollon oil spill, Florida. *Photo courtesy of St. Petersburg Times and Evening Independent.*

The contamination affected thousands of birds in the area, mostly loons, cormorants, mergansers, and scaup, while marine life in the intertidal zone was also harmed by the

oil. Hundreds of students from local high schools and Florida Presbyterian College volunteered to help clean the birds. The Audubon Society also formed a kind of adoption center where volunteers took birds home to care for them until they appeared able to return to the shores. Unfortunately, since many of the cleaned birds had ingested oil while preening their feathers, they died despite these efforts.

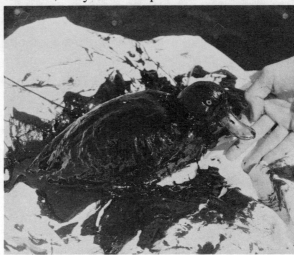

Delian Apollon oil spill, Florida, U.S.A. *Photograph courtesy of St. Petersburg Times and Evening Independent.*

18. *OCEANIC GRANDEUR* OIL SPILL
Torres Strait, Australia   **March 3**

The Liberian tanker *Oceanic Grandeur*, carrying 55,000 tons of crude oil, struck a submerged object 20 miles off the coast of Thursday Island in the Torres Strait, rupturing five of the ship's carrier tanks. Although the leakage was stopped one day after the accident, several oil slicks were reported. Collectively, the slicks covered an area about 30 miles long and 15 miles wide, which floated eastward toward the Barrier Reef, some 200 miles away from the point of discharge.

Immediate steps were taken by the government to prevent the slick from spreading; and most of the oil was reported dispersed within days after the clean-up operation. However, some marine life, especially a "farm" of cultured pearls, was reported injured by the oil.

19. CALINGIRI EARTHQUAKE
Western Australia   **March 10**

An earthquake, with its epicenter near Calingiri, 60 miles northeast of Perth, created a three-mile ground fault after producing a tremor felt 120 miles away. The quake registered magnitude 6.0 on the Richter scale. No aftershocks were reported.

20. BRETON SOUND OIL SPILL
Louisiana   **March 10**

The dynamite blast used to extinguish a month-long oil fire on the Chevron Oil Company's Platform 11 off the Louisiana Coast also created a 100-foot-high fountain of crude oil. The oil began affecting the offshore seafood beds almost immediately.

Chevron installed a pollution control system to contain and clean up the oil flowing from the well at the rate of 600 to 1,000 barrels a day following the blast. The 85-square-mile slick spread in a north-northwest direction toward Breton Island. The chemicals used to contain the slick caused great concern among conservationists, for it caused the oil to sink to the ocean bottom where it could damage marine life and even deplete the water of dissolved oxygen.

By March 20, the slick had shrunk to some 20 square miles, partly because of wave action. Even though the oil disappeared, conservation officials feared Louisiana oyster beds, fisheries, and coastal wildlife had been seriously harmed by the spill.

21. *OTHELLO* OIL SPILL
Sweden   **March 20**

The oil tanker *Othello* collided with the tanker *Katelysia* in a ship channel between Sweden and Finland. Although both ships could move under their own power after the collision, about 60 to 100 metric tons of oil were spilled from the *Othello* into Tralhavet Bay, east of Vaxholm, Sweden, a resort area north of Stockholm.

Part of the oil disappeared beneath two-foot-thick ice that lined both sides of the ship channel. Because of the water temperature, most of the oil formed large blobs about two feet in diameter. The blobs of oil, except for the few inches appearing above the surface, sank in the icy water. A new chemical, Cab-O-Sil St-2-0, was used to burn away the blobs. Because of the successful use of the agent, no damage to wildlife was reported.

22. GUJARAT EARTHQUAKE
India   **March 23**

Twenty-four people were killed and over 200 injured and at least 3,500 houses were damaged and 13,000 people left homeless after an earthquake struck central Gujarat, India.

The quake's epicenter was located 25 miles from the city of Broach, where extensive damage was reported. The quake registered magnitude 5.7 on the Richter scale. There were no reports of fissuring or landslides.

23. WESTERN TURKEY EARTHQUAKE
Turkey   **March 28**

On March 28, western Turkey was shaken by a severe earthquake of Richter magnitude 7.1. The quake occurred in the upper reaches of the Gediz River and affected the region between the towns of Kutahya, Emet, Simav, and Usak. The area previously suffered devastating shocks in 1866, 1896, 1930, 1943, and 1944.

In the province of Kutahya, 1,000 people were killed, 520 were injured, and over 80,000 were left homeless. In an area of 3,000 square kilometers, 3,500 housing units collapsed, 7,700 were badly damaged, and another 10,000 suffered moderate damage. More than 1,000 public buildings were badly damaged, 240 beyond repair. The cost of damage was estimated at about $21.6 million.

Western Turkey earthquake, Turkey.

24. CHAPARRASPIQUE VOLCANIC ACTIVITY
El Salvador   **March 30**

Ash erupted from the crater of the San Miguel Volcano in El Salvador three times on the morning of March 30. The first eruption—accompanied by a loud explosion heard several miles away—sent ash particles flying some 1,200 feet above the crater floor.

The ash fell in a fan-shaped area extending at least six miles northwest of the summit. Fortunately, the city of San Miguel, six miles northeast of the peak, remained untouched. Eruptions of ash continued into early April.

25. GUAJAKI TRIBE DISCOVERY
Paranambu, Paraguay   March 30

A group of "stone-age" nomadic indians was discovered in Paraguay. The tribe was identified as the Guajaki, a people whose culture and perhaps very existence seemed endangered by the expansion of logging and agricultural enterprise in the area.

The tribe was discovered about 25 miles west of Paranambu on the Rio Parana during a forest inventory carried out by the Food and Agriculture Organization of the United Nations and the Paraguay Ministry of Agriculture. The inventory crew returned artifacts from the tribe's village, which had been abandoned only shortly before the crew's arrival.

An ethnologist specializing in the Guajaki said he believes the members of the tribe have probably never seen a European. He also said that the current economic expansion of the country may seriously affect the Indian culture, since only these few Guajakis still live in their natural state in the region.

26. SOUTHERN MINDANAO EARTHQUAKE
Philippine Islands   March 30

An earthquake of Richter magnitude 7.0 was recorded with an epicenter approximately 50 miles south of the city of Davao. The tremor was reported as very similar to the Davao earthquake of January 10.

27. MOROCCAN GRAIN CROP DESTRUCTION
Morocco   March

During the last week of March, wheat and barley fields were attacked by Spanish field sparrows (*Passer hispaniolensis*). Damage to the early maturing wheat plants varied from 5 to 50 percent. At a research station near Tamellait, an entire 40-acre field of barley was destroyed by the birds. Since the birds migrate from Spain to Morocco, an international conference to study the control of the sparrow was established. The government of Egypt had previously reported a serious sparrow infestation in 1969.

28. LAKE ST. CLAIR AND LAKE ERIE MERCURY CONTAMINATION
U.S.A.—Canada   March–May

In the spring of 1970, all fish in Lake St. Clair—regardless of species or section of lake—were found to exceed the acceptable mercury level (0.5 parts per million). Because of the high level, the Canadian and American governments closed the lake to sport and commercial fishing. On April 29, the state of Michigan announced a similar embargo. In the Ohio waters of Lake Erie, however, an embargo was placed only on wall-eyed pike.

Normally, there is a zero tolerance level for mercury residue in edible fish, but the guideline in use by the U.S. Food and Drug Administration is now 0.5 ppm for the wet, edible tissue of fish.

The mercury level in wall-eyed pike in Lake St. Clair reached 5 ppm. In other fish of a less predatory species, such as perch, levels as high as 2 ppm were still reported. Wall-eyed pike in the western basin of Lake Erie, where mercury contamination was lower, still averaged over the 0.15 limit. Mercury levels in Lake Erie perch were approximately 0.5 ppm; and generally lower among other species.

29. GUAYAS PROVINCE FLOODING
Ecuador   April

Heavy rains during the first half of April caused the Daules and Vinces rivers to overflow their banks, flooding the flat, low-lying plains of Guayas Province in Ecuador. About half of the flooded area was cultivated, mostly rice, corn, tobacco, and cotton.

30. MASAYA VOLCANIC ACTIVITY
Nicaragua   April 4

A small eruption from the central vent of an old lava lake at the Santiago crater of Masaya was observed in early April. Two major explosions were heard during the two days of activity. Several minor explosions and strong fumarolic activity from the central vent were also observed.

31. LUZON EARTHQUAKE
Philippine Islands   April 7

A strong earthquake (Richter magnitude 7.2) struck the island of Luzon 100 miles northeast of Manila. The tremor created a massive landslide and fissures including one that was half

a mile long and two feet wide. Trees were uprooted and water and sand shot out of the fissures. There were also reports of a "glowing brightness" atop the surrounding mountains. During the quake a deafening sound, "like the noise of many low-flying jets," was also reported.

The earthquake was apparently responsible for the deaths of at least five persons and for considerable property damage.

32. GULF OF ALASKA EARTHQUAKES
Alaska   **April 16**

A series of earthquakes occurred in the Gulf of Alaska on April 16, 18, and 19. The tremors were located 250 miles southeast of Anchorage. The tremors registered a Richter scale magnitude 6.6 on April 16, 6.3 on April 18, 6.0 on April 19. The second tremor was widely felt and caused more damage in Anchorage.

33. TARUT BAY OIL SPILL
Saudi Arabia   **April 20**

A large pipeline carrying crude oil to a refinery owned by the Arabian American Oil Company broke during a storm on April 20 and spilled an estimated 100,000 barrels of oil into Tarut Bay off the eastern coast of Saudi Arabia. Because of favorable wind conditions and effective methods of oil containment and clean-up, none of the oil entered the Persian Gulf.

Booming, skimming, and use of Corexit helped contain and disperse the oil. The high air and water temperatures and the high salinity of the water were thought to have contributed to a rapid rate of biodegradation of the oil.

Many coastal areas, however, were affected by the oil. Where heavy slicks had passed in and out with the tides for several days, the fauna was largely killed by direct contact with the oil. The mortality included crabs, bivalves, and limpets. At the edge of a heavy slick, numerous dead fish were washed ashore. Oil had also blackened exposed patches of coral beachrock, a shelf that underlies the sand or mud of the bay. There was, however, no tainting of fish caught in the bay during clean-up operations.

Tarut Bay and Tarut Island are largely surrounded by extensive marshes of dwarf mangroves and other saltmarsh plants. These plants literally mopped up considerable amounts of oil. Mangrove leaves were oiled on the lower half of the shrub, but the soil level was usually free from oil. Three months later, some plants had lost their leaves, but many survived, some bearing flowers and fruits.

34. MOUNT ASO VOLCANIC ERUPTION
Japan   **April 21**

Mount Aso became active for the first time in four years, spewing volcanic ash and gases some 500 feet into the air. The eruption came from a new opening inside the first crater on the Nakadake volcanic cone of Mt. Aso.

Initially on April 21, a small amount of ash erupted from a crack in the north-northeast corner of the crater's bottom. On April 22, a depression about 60 feet long and 30 feet wide was observed in the same place where the new crater was formed; it spewed gray ash into the air reaching the brim of the outer crater 450 feet above it. An eruption of this size was the first of its kind at Mt. Aso since May 1966.

35. MOUNT MALINAO VOLCANIC ACTIVITY
Luzon, Philippine Islands   **April 24**

Earth tremors and deep rumblings were reported at Mt. Malinao, a supposedly extinct volcano 10 miles northwest of Mt. Mayon. No subsequent volcanic activity of any significance was reported.

36. ALASKA PENINSULA OIL SPILL
Alaska   **April 25**

A slick of what appeared to be light diesel oil was thought to have devastated wildlife along a 400-mile stretch of the western coast of the Alaska peninsula. Aerial surveys of peninsula beaches on April 27 showed an estimated 86,000 dead birds, mostly common murres.

The diesel oil was thought to have come from two Japanese ships that sank a week before.

However, the bird mortality may have been unrelated to the oil spill, as beach surveys failed to identify any oil or oily substance associated with the waterfowl mortality. Laboratory analysis failed to identify any hydrocarbons internally or externally on a sample of birds collected by field investigators. Apparently still weak after wintering in the Bering Sea, the murres more likely died from a combination of starvation and exhaustion, a condition intensified by a severe storm that prevented them from feeding for several days.

Some mammals, however, including killer whales, were reportedly affected by the oil. About 400 seals in the area were observed acting in a strange manner (they refused to enter the water), and having a white, glazed look in their eyes. The shores were also lined with dead starfish, scallops, and finfish, presumably killed by the diesel slick.

37. GUATEMALA BASIN EARTHQUAKE
Guatemala   April 29

A doublet shock—two earth tremors only 15 seconds apart—occurred in the Guatemala Basin approximately 175 miles southwest of Guatemala City. The twin quake registered Richter magnitude 7.5. No damage, casualties, or fissuring were reported, although 70 seismic waves were recorded in nearby El Salvador. In nearly the same area, a stronger earthquake (magnitude 7.8) occurred on July 28, 1957, causing 55 deaths and extensive property damage.

38. *POLYCOMMANDER* OIL SPILL
Western Spain   May 5

The Norwegian oil tanker *Polycommander* caught fire in the mouth of the River Vigo and ran aground on the east coast of the Cies Islands. The tanker was carrying 50,000 metric tons of crude oil of a light type with a high proportion of volatile components.

The fire was extinguished on May 6 after an estimated 10,000 metric tons of oil spilled into the sea. About half of the spilled oil washed ashore along the coastal areas at the entrance to the River Vigo. Canvas sheets were used successfully to contain the spill, and absorbent powders, such as bentonite and china clay, were spread on slicks to sink the oil. Except in areas where there was mussel cultivation, Corexit was used with success. No mortality of fish or birds was observed, and tests showed the mussels to be free of contamination. Beaches were cleaned by pumping emulsified oil into tanks, either by dispersants or mechanical means.

39. CROOKED CREEK FISH KILL
Missouri   May 5

An estimated 349,000 fish died in Crooked Creek following the dumping of considerable, but undetermined, quantities of toxic material into the water. Samples taken from the creek showed the material to be a solution of chlordane and Malathion in xylene. The Malathion was thought responsible for the majority of the fish killed.

About 90 percent of the fish killed were minnows and orange-throat divers. In addition to fish, all other aquatic life in the stream was killed for a distance of two miles downstream from the dumping, including snakes and turtles, plus large numbers of frogs, tadpoles, and crayfish.

40. ÄLVSBYN REINDEER MORTALITY
Northern Sweden   April–May

A herd of reindeer owned and tended by one Lapp family was nearly destroyed when the animals ate leaves of trees sprayed with herbicides. In early April, a short period of mild weather was followed by a severe cold spell and the reindeer, unable to move through the snow to obtain food, apparently ate some cultivated plants sprayed with herbicides.

Within days, some 250 animals died or disappeared; 100 corpses were found. (In a herd of 600, usually no more than 25 to 30 animals disappear during the winter stay in the woodlands. Indeed, most are later found in other herds and brought back to their owners the next fall.) The death rate ceased immediately when the animals were given fodder brought in from other areas.

41. SAKURAZIMA VOLCANIC EXPLOSION
Japan   May

A lava mass about 300 feet wide appeared in the summit crater of the Sakurazima Volcano at the beginning of May, without smoke emission and volcanic earthquakes. On June 12, however, the crater exploded for the first time since formation of the lava mass. Other explosions—with accompanying airshocks, smoke, and flames—continued throughout the year and formed a new crater at the Minamidake south peak. During 1970, Sakurazima exploded 19 times.

42. HEKLA VOLCANIC ERUPTION
Iceland   May 5–July 5

The main eruption of the Hekla Volcano occurred along fissures running from the base of the Hekla ridge to both the southwest (Sudurgigar) and northeast (Hlidargigar). During the first phase of eruption, clouds reached as high as 45,000 feet. The total volume of tephrite produced was about 70 million cubic meters. This material, with high sulphur dioxide and fluorine content, was carried northwest. Fallout was observed along the north coast of Iceland, 50 miles from Hekla, 2.5 hours after the eruption. Because of the high fluorine content of the tephra, some 7,500 sheep died.

Lava flow from the Sudurgigar fissure ceased on May 10 and from the Hlidargigar craters on May 20. On that same day, however, a new half-mile-long fissure opened north of Hlidargigar. Lava flowed from this fissure until July 5.

43. KARIMSKY VOLCANIC ERUPTION
Kamchatka, U.S.S.R.   May 11

After lying dormant for three years, the Karimsky Volcano erupted, destroying the rigid upper part of the magmatic column and creating a crater 600 feet wide on the summit.

Karimsky volcanic eruption, Kamchatka, U.S.S.R. One of the explosions which occurred in May of 1970. *Photograph courtesy of N. V. Ogorodov, Institute of Volcanology, Petropavlovsk Kamchatsky, U.S.S.R.*

The frequency and power of the volcanic outbursts increased sharply on May 12, until an eruptive cloud reached a height over four miles. The entire volcano region was covered with pumice and ash fall. Also, incandescent avalanches were reported along the western and southern slopes of the cone.

An outflow of lava began on May 13 and eventually covered the northeastern sector of the volcano slope to a thickness of about 15 feet. After May 15, explosive activity became continuous with outbursts following each other almost every two hours for two weeks. Later, the intervals between explosions increased to two or three days.

44. HOLMES FIREBALL
Colorado–Wyoming   May 18

A large fireball seen over northern Colorado and southern Wyoming apparently "crumbled and burst" at a point south of Foxpark and over the town of Holmes, Wyo. One large fragment—between one-fifth to one-eighth of the total object—continued to be seen in flight after the explosion. The fireball was generally described as an "intense white," resembling, according to one person, "a magnesium flare," or as another put it: "a Fourth of July skyrocket in flight." Some observers reported a spiral motion for the fireball.

Because of the heavily forested and largely inaccessible terrain of the Wyoming–Colorado area, no field research party was dispatched to the scene.

45. SUWANOSEZIMA VOLCANIC EXPLOSION
Japan   May 28

Ostake, the summit crater of Suwanosezima Volcano, exploded on May 28 and again on September 30, producing spurts of ash and cinders every four or five minutes. Ostake exploded again in December, this time sending volcanic smoke 3,000 feet in the air. No damage was reported.

46. HUARAZ EARTHQUAKE
Huaraz, Peru   May 31

A violent earthquake with an estimated Richter magnitude of 7.5 was recorded some 30 miles off the coast of the city of Chimbote, Peru, 200 miles northeast of Lima.

Huaraz earthquake, Huaraz, Peru. A cemetery in Caraz where dozens of coffins were thrown out from niches. *Photo courtesy of La Prensa. Lima, Peru.*

The earthquake, perhaps the most destructive in Latin American history, killed more than 50,000 people, principally as a result of building collapses. Almost 20,000 people disappeared in an avalanche that buried the city of Yungay and the village of Ranrahirca.

The avalanche, which occurred within only minutes after the quake, followed the Santa River's course, covering a great part of the valley with thick deposits of mud and destroying a great many bridges, communications systems, and cultivated fields. The avalanche also caused the level of the Santa to rise and to generate a huge wave—some 45 feet high—that rushed downstream through the river's narrow canyon destroying everything in its path.

47. SOUTHEASTERN EUROPE FLOODING
Europe   **May–June**

Abundant snowfalls and torrential rains during April and the first half of May caused all the rivers in Rumania to overflow their banks and to flood much of that country.

The floods covered areas cultivated with cereals, vegetables, and other crops. Some 40,750 homes were destroyed or damaged, and some 8,092 other buildings were destroyed. Approximately 268,000 people were evacuated from the flooded areas, but some 35,000 farm animals and some 61,000 fowls drowned. The catastrophic waters destroyed or damaged 624 major bridges, 1,518 smaller bridges, 250 miles of asphalt highways, 1,000 miles of cobbled roads, 1,300 miles of forest roads, 280 miles of railway lines, and 1,000 miles of electric lines.

Subsequent earth slides caused by the huge rainfalls and flooding damaged hundreds of other buildings, and completely destroyed one small village.

48. IMMINENT KOUPREY EXTINCTION
Cambodia   **June–August**

The Cambodian kouprey, a "living fossil" with only some 70 specimens known in the world, was threatened with imminent extinction due to war action near its native habitat in a remote area of Cambodia.

Huaraz earthquake. Huaraz. Peru. Local traffic on severely damaged highway ruined by the tremor. *Photo courtesy of La Prensa. Lima. Peru.*

The kouprey differs so greatly from other bovids that it is considered a separate genus. The animal is extremely primitive in many of its anatomical features and is very likely a close relative to the missing ancestors of the Indian Brahmin, or hump, cattle.

The Cambodian government sent an expedition into the area to capture three or four kouprey as the basis for a captive herd. Unfortunately, the animals proved to be allergic to the nicotine drugs used as tranquilizers and none lived more than 14 hours in captivity.

Disturbing reports indicated that the area containing the last kouprey herds had been overrun with Vietcong, and the herds might have been slaughtered for food. No kouprey exist in captivity anywhere in the world.

49. NORTHEASTERN CATERPILLAR INFESTATION
Northeastern U.S.A. **June–August**

An infestation of the saddled prominent caterpillar, *Heterocampa guttavitta,* has been reported in the states of Massachusetts, New York, Vermont, New Hampshire, and Maine during the last three years, 1968–1970. The saddled prominent is a leaf-eating caterpillar that defoliates mostly sugar maple, beech, and birch trees.

To a lesser extent, the green striped maple worm, *Anisota rubicunda,* and the variable oak leaf caterpillar, *Heterocampa manteo,* were part of the infestation.

In the summers of 1969 and 1970, the area of Massachusetts defoliated by the caterpillar increased to 100,000 acres and spread from the Berkshires to Franklin and Hampshire counties. Although the Massachusetts Department of Natural Resources had not initiated the use of insecticides to combat the caterpillars, private landowners sprayed infested areas with Sevin, a carbamate pesticide. Since the caterpillar undergoes cyclic increases in population, the infestation was expected to decline, perhaps with the aid of such natural predators as parasitic wasps, the calasoma beetle, and a pathogenic virus.

50. SINO–SOVIET BORDER EARTHQUAKE
U.S.S.R.–People's Republic of China June 5

About 5,000 houses were destroyed, leaving some 20,000 people homeless, after an earthquake of Richter magnitude 6.8 struck the Chinese-Russian border region. Strong underground shocks continued in the area until June 8.

51. NAINI TAL SUDDEN SKY BRIGHTENING
Uttar Pradesh, India **June 9**

On a moonless and cloudless night, two Indian astronomers at a Smithsonian-affiliated observatory noticed a sudden and extraordinary sky brightening. "For about one second," one scientsts said, "the total local sky suddenly brightened up to an intensity comparable to that of the Milky Way during these summer months." According to the report, the event took place in two stages: first, the sky became bright; then, immediately thereafter, the brightening intensified momentarily. No explanation of the phenomenon was given.

52. WALAGA METEORITE FALL
Southwest Ethiopia **June 19**

Three meteoritic specimens were recovered near the village of Jarso, in the province of Walaga. The stones, apparently all part of an original single object that broke into three pieces, weighed 2.3 kilograms, 139 grams, and 17 grams respectively.

The fall had been accompanied by auditory phenomena.

53. QUEEN CHARLOTTE ISLANDS EARTHQUAKE
British Columbia, Canada **June 24**

An earthquake shook a widespread area of northern British Columbia on June 24. The quake, recorded at Richter magnitude 7.0, with an epicenter near the Queen Charlotte Islands, was preceded by a foreshock at 5.8 magnitude. There were no reports of damage or injuries.

54. CHELDAG MUD VOLCANO ACTIVITY
U.S.S.R. **June**

The mud volcano Cheldag, located 40 miles west of Baku, awakened after a dormant period lasting 100 years. This awakening was marked by a powerful emission of gas. More extraordinary, a cupola-shaped hill of cementlike fragments of dry clay gradually formed in the center of the volcano's crater. The mud dome reached 30 feet in height and about 180 feet in diameter.

In July, a second island, measuring 300 feet long, 150 feet wide, and four feet high, appeared in the Caspian Sea as a result of an underwater mud volcano.

55. DAMGHAN EARTHQUAKE
Iran   **July 30**

Four earth tremors, the largest registering 6.3 on the Richter scale, were responsible for the deaths of 40 people. Collectively, the tremors represented the strongest earthquake activity to hit Iran since August 1968.

56. KONERT SPRING GAS LEAK
Missouri   **July–August**

Poisonous gas rising from a natural sinkhole leading to an underground channel called Konert Spring apparently was responsible for the deaths of many wild animals beginning in mid-July. Several opossums and skunks and about 35 birds were found dead in the vicinity of the sinkhole. Large numbers of insects apparently also died from the gas after attempting to lay eggs in the decomposing animal flesh.

The spring is located in Jefferson County, near St. Louis. Tests showed the gas contained up to 3 percent methane.

Although the actual source of the gas was not determined, two possibilities were considered likely: (1) the decomposition of naturally accumulated organic material producing marsh gas, or (2) the decomposition of sewage and industrial waste. After further investigation, the latter theory was believed to be the most probable. The gas, mostly nitrogen and carbon dioxide, contained small amounts of oxygen, carbon monoxide, cyanides, sulphates, and hydrogen sulphide.

57. AMAZONAS EARTHQUAKE
Amazonas, Colombia   **July 31**

The first earthquake reported in the Amazonas area near the Peruvian border in the last ten years was felt in such widely separated locations as Bogota, Lima, Caracas, and Buenos Aires on July 31. Fortunately, the quake, recorded at Richter magnitude 7.0, was confined to the jungle and caused little damage elsewhere in South America.

58. NORTH HALEDON FIREBALL
New Jersey   **August 2**

An object seen over New Jersey and de-scribed as a "reddish ball of fire about one half the size of the moon" turned out to be something less exotic. The fireball was actually an aerial balloon with a flare (or some other sort of slow-burning spark-emitting device) attached to it.

59. MOZAMBIQUE CHANNEL FIREBALL
Indian Ocean   **August 6**

An observer aboard a boat sailing through the Mozambique Channel reported an extremely bright fireball on August 6. The object apparently exploded, creating a bright flash before particles descended slowly beyond the horizon.

60. OHIO FIREBALL
Ohio   **August 8**

An extremely bright fireball, reported to be much brighter than a full moon, traveled across the skies over the state of Ohio. One observer said the object was "a bluish-white flash which lasted two or three seconds." Another called it "a greenish flash, with red streaks and sparks falling away from it." Observations of the meteor were also made over West Virginia and Tennessee.

61. MALAKAL METEORITE FALL
The Sudan   **August 10**

A two-kilogram meteorite was recovered near Malakal in the Upper Nile region of the Sudan after a fall was observed on August 10.

The outside of the rock was black, with a smooth varnished appearance. The interior was greenish-gray, brecciated and criss-crossed by numerous small fractures. It was generally composed of silicates (80 to 85 percent olivine and pyroxene) and metallic iron (about 15 to 20 percent). The meteorite was classified as an olivine hypersthene chondrite.

Malakal meteorite fall. The Sudan. *Photograph courtesy of J. R. Vail. Dept. of Geology. University of Khartoum. Khartoum. Sudan.*

62. NEW HEBRIDES EARTHQUAKE
New Hebrides, South Pacific   August 11

A major earthquake, at Richter magnitude 7.6, occurred in the region of the New Hebrides Islands. A foreshock of magnitude 6.8 was recorded the previous day. The quake caused only minimal damage.

63. DECEPTION ISLAND VOLCANIC ERUPTION
Antarctica   August 12

A volcanic eruption at Deception Island was recorded by seismographs on August 12. Because of winter conditions, however, it was impossible for scientists at Antarctic bases to fly over the island. There were also reports of a large ash rain caused by the volcano.

64. LAKE LINDA DRAINAGE
Juneau, Alaska   August 13–15

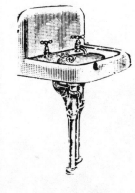

Superglacial lakes, such as Lake Linda in Alaska, often disappear in a flash. When triggered by an avalanche or earthquake, the sudden draining of such superglacial lakes can sometimes cause catastrophic flooding. Oddly enough, between August 13 and 15, Lake Linda drained very slowly, almost imperceptibly, without any outward effects.

On August 13, the water level of the lake dropped some three to four inches per hour. The lake, filled with icebergs, did not produce any perceptible current by its drainage. An obvious connection between the drainage and Lynn Falls did not prove to be valid. By August 15, no water could be seen among the huge blocks of cracked icebergs, and it was assumed that the lake had completely drained.

65. RINCÓN DE LA VIEJA VOLCANIC ACTIVITY
Costa Rica   August 14–25

A new eruption of the volatile Rincón de la Vieja Volcano August 14 was characterized by great clouds of ashes and smoke. On August 25, the volcano erupted again, this time producing only a minor ash cloud.

66. LAKE ERIE MUD PUPPY MORTALITY
Canada–U.S.A.   August 16

Over 50 dead mud puppies (salamanders) were found on a stretch of a Lake Erie shoreline west of Buffalo, N.Y. Counts of significant numbers of dead salamanders were found for 20 miles along both the Canadian and American shores of the lake.

From their slightly bloated appearance, the salamanders were thought to have been dead for several days. Laboratory examinations were made to determine if the cause of death was due to chemicals, sewage, disease, mercury, lack of oxygen, increase in water temperature, an algae bloom, or some other factor. Despite the extensive tests, the cause of the mud puppy mortality was never determined.

67. TELICA VOLCANIC ERUPTION
Leon, Nicaragua   August 17–25

The Telica Volcano began a new eruption on August 18 with small, sporadic eruptive emissions. But on August 18, the volcano began emitting columns of very dark, almost black ash. Although heavier particles fell near the crater, ash particles remained suspended in the atmosphere and were carried by air currents toward the Pacific coast. The clouds dumped ashes on the cities of Chichigalpa, Posoltega, and Corinto, and affected cotton crops in the region.

68. STUDLEY POND FISH KILL
Rockland, Massachusetts   August 19

Studley Pond was treated on August 19 with copper sulfate, Diuron, and Silvex to destroy the weeds, algae, and duckweed that blanketed the pond. One week later, the bodies of several thousand fish floated to the pond's surface. Although there was evidence of pollution from industrial wastes and from a nearby sewage treatment plant, the fish kill actually seemed the result of lowered oxygen levels due to the decomposition of the pond's plant life.

69. STUTTGART FIREBALL
Stuttgart, Germany   August 27

Ten stations of the European All-Sky Camera Network photographed a fireball of intense brightness that entered the earth's atmosphere over Stuttgart. A search for meteorites was organized by the Max-Planck-Institut für Kernphysik, which operates the stations, but no materials were found.

70. VIRGINIA BEACH FISH KILL
Virginia Beach, Virginia   August 31

An estimated 20,000 to 50,000 fish, as well

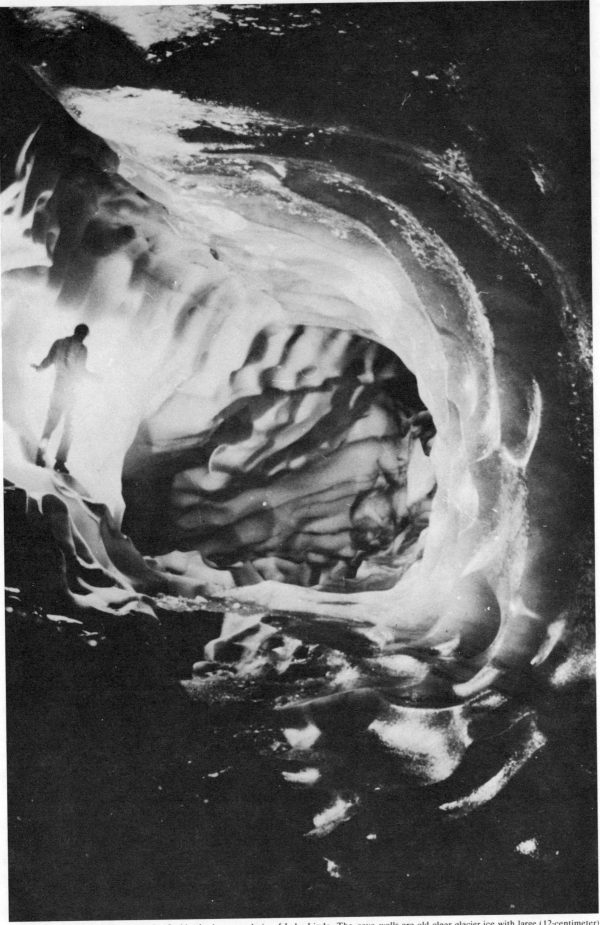

Lake Linda drainage, Juneau, Alaska. Inside the ice cave drain of Lake Linda. The cave walls are old clear glacier ice with large (12-centimeter) crystals. *Photograph by Robert Asher. Geological and Arctic Sciences Institute. Michigan State University. U.S.A.*

as other forms of marine life, were killed in the inshore waters along a seven-mile stretch of beach near the Back Bay National Wildlife Refuge at Sandbridge, Va. All species of both vertebrates and invertebrates indigenous to the shore area were killed, including surf clams, blue crabs, barnacles, starfish, flounder, spot, sea trout, and other game and food fish.

Oddly enough, the fish kill occurred only in a remote area where no sewage or industrial waste was being released. A nearby Navy antiaircraft station said there had been no violation of the restricted area. One theory suggested that a container of highly toxic material struck the beach's soapstone and ruptured, releasing its deadly contents into the area.

Tests were made for pesticides and poisons, but they proved negative. The cause of the kill was undetermined.

71. WOLENCHITI FRACTURING
Ethiopia   August

Severe fracturing of the ground along the main Ethiopia rift appeared near Wolenchiti, some 50 miles southeast of Addis Ababa. The fractures were reported to be 18 feet deep. The fissures were similar to those that appeared 10 miles south of Wolenchiti in August 1966.

72. MONARCH BUTTERFLY MIGRATION
North America   August–October

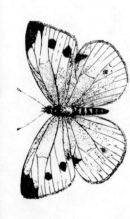

During the early fall, literally millions of giant Monarch butterflies, *Danaus plexippus,* were observed migrating from the northern midwest United States and southern Canada to the Gulf States and Mexico. Large migrations of Monarchs usually occur every six years, but one this size had not been witnessed in recent memory.

Fluctuations in the Monarch's population level are believed to be related to a virus that infects the insect. During peak years of population, the butterflies seem immune to the disease. In other years, however, prevention of infection is impossible even under the most sterile laboratory conditions. The Monarch migrations usually start in mid-August and last through October, depending on the temperatures. Since migrations usually start in the summer when temperatures are still high and food still abundant, there may be other

factors than the virus involved in triggering the mass movement.

Monarch butterfly migration, North America. Massed group of Monarchs resting on a grape vine in Point Pelee, Ontario, on the north shore of Lake Erie.

73. SOUTHERN CORN LEAF BLIGHT
U.S.A.   August

Southern corn leaf blight, *Helminthosporium maydis,* a well-known pathogenic fungus that attacks corn, historically has been restricted to the southeastern part of the United States, and had not previously caused serious damage to corn production.

In the summer of 1969, however, a more virulent strain of the blight was discovered in Illinois. During the following winter, it was again observed in winter plantings in Florida. At that time, the fungus was not causing serious damage, but by the spring of 1970, the fungus had caused serious damage to corn crops in Florida, Georgia, Alabama, Mississippi, and Texas. As the growing season progressed, the blight moved north to Tennessee, Kentucky, Ohio, Indiana, Illinois, and Iowa. As the growth pattern of the fungus unfolded, it became clear that it was a new biotype that presumably arose from a mutation or some other genetic change.

A vital characteristic of the new form of fungus is that it infects more of the corn plant than its predecessor, which only grows in the leaf of the plant. The new strain attacks the leaf, the stem, and the ear, and therefore possesses the potential to inflict much greater damage to the plant.

74. GULF OF MEXICO FISH KILL
Southern Gulf of Mexico   August–Sept.

Mexican shores were covered with the bodies of fishes of different sizes and different species—including a shark—during the last days of August. Divers in the Bay of Campeche found that the Gulf bottom was also covered with large quantities of dead fishes. Although no reason was given for the kill, local people said that the phenomenon occurred almost every year at the same location and at the same season—at the end of heavy rains.

75. ESCAMBIA BAY MENHADEN KILL
Escambia Bay, Florida   September 3

The September fish kill in Escambia Bay, involving the deaths of 10 to 15 million menhaden, was actually one of 42 fish kills that occurred there since June 21. The menhaden kills had occurred every year for the last ten years due to overnutrification; the extent of the kills varied from hundreds to millions.

Most of the kills occurred in the small bays off Escambia Bay. Algae and plankton populations grew rapidly here because of overnutrification resulting from industrial discharge into the water. When the menhaden came in to feed on the plankton, there was insufficient oxygen for the fish.

The most extensive kill occurred in July 1970 in the Escambia River where all fish species died. Oxygen deficiency was thought to be the most plausible reason for the mortality.

Menhaden are used for food, oil, livestock feed, fertilizer, and are a source of food for other commercial fish. The industries involved were ordered to take corrective measures by January 1973. Because the present conditions are the result of 10 to 15 years of misuse, it is expected that even with the new order, it

would take at least six years before any positive results were discernible.

76. LAKE MILLS FIREBALL
Wisconsin   September 10

A bright light green fireball was reported traveling west to east over Wisconsin. The object fragmented into three separate pieces before descending beyond the horizon.

77. DUNAFÖLDVÁR LANDSLIDE
Dunaföldvár, Hungary   September 15

Dunaföldvár, a small town near the Danube River, located 40 miles south of Budapest, was the scene of a great landslide, in which more than one million cubic meters of earth slid down into the river accompanied by an intense rumbling sound. Due to the movement, a great deal of the high bank near the Dunaföldvár railway bridge was squeezed into the water, while an island "mountain chain" formed in the river.

Dunaföldvár landslide. Dunaföldvár. Hungary.

A vineyard, a fruit garden, two small houses, and a country house were moved downward about 90 feet, but only one of the houses collapsed. No deaths and only slight injuries to area inhabitants were reported.

As a consequence of the slide, some considerable soil deformations were reported. In a vineyard there were fissures parallel with each other and parallel with the riverside. In other parts of the bank, 20-foot-deep chasms were created. Most extraordinary, however, were the islands. Immediately after the landslide, an enormous cloud of dust rose into the

Dunaföldvár landslide, Dunaföldvár, Hungary.

air. When the cloud cleared, observers could see two new islands in the river each about 900 feet long and 25 feet high.

The surfaces of the two long and narrow islands are extremely rough and contained deep fissures. In some places, there were tilted blocks of clay taller than a man. Because so many characteristically Danubian shells and snails were found on the islands' surface, the long islands were assumed to have emerged from under the water and not from the riverside.

The landslide was most likely caused by the collapse of a subterranean cavity, although one of the contributing causes to the slide could have been abundant rains in the summer. The land slipped downhill to the river bank where its accumulated weight pushed up the river floor to form the islands.

78. RIO SOLIMOES FISH KILL
Rio Solimoes, Brazil   September 16

A major fish kill, involving mostly mapara, *Hypopthalmus edentatus,* was related to the entrance of a cold air mass into the Amazonia area.

79. BEERENBERG VOLCANIC ERUPTION
Jan Mayen Island, Greenland Sea   Sept. 20

Four large fissures on the east side of the 6,800-foot Beerenberg mountain opened up after a volcanic eruption September 20. Basalt lava flows from all four fissures reached the sea; and smoke and ashes were reported reaching heights of 18,000 feet and were visible 100 nautical miles away.

Thirty-nine persons on Jan Mayen Island were evacuated by plane the evening of the eruption.

The volcano was so large it could be photographed from the ESSA 8 satellite.

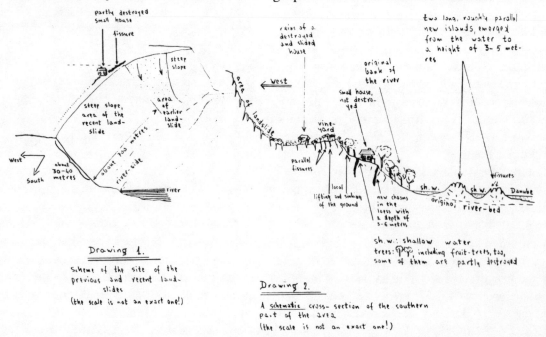

Dunaföldvár landslide, Dunaföldvár, Hungary. *Drawing detailing the Dunaföldvár landslide by Dr. Peter Hedervari. International Association of Planetology. Budapest, Hungary.*

80. COLORADO–NEBRASKA FIREBALL
U.S.A.   **September 21**

A bluish-white fireball as bright as the full moon was reported from the town of Wray, Colo., near the Nebraska–Kansas border. The very concentrated light lasted for about two seconds, producing a momentary vapor trail. No explosions or sonic booms were reported.

81. HOLSTON RIVER MERCURY CONTAMINATION
Smyth County, Virginia   **September 22**

Fishing in the north fork of the Holston River was prohibited by the state's health commission after the discovery of high levels of mercury in fish samples. Levels eight times the allowed limit were found in the edible portions of Holston River fish.

A chemical plant in North Saltville, Va., was thought to have been responsible for the contamination. Because of the persistence of mercury—some of which was absorbed by the algae—the fishing season was closed for the year.

82. TANZANIA ARMYWORM OUTBREAK
East Africa   **September**

The armyworm is a velvety black caterpillar about 1.25 inches long that has been attacking cultivated crops in Tanzania and other Eastern African countries for more than a decade. Because of lack of knowledge concerning pesticides, and because the worms were not iden-

Tanzania armyworm outbreak. East Africa.

tified by farmers during the early stages, the 1970–1971 outbreak was the heaviest and most destructive since 1961.

A female caterpillar can lay as many as 1,000 eggs which hatch in about three days' time. Once hatched, the larvae are difficult to detect until fully grown. The whole cycle begins again when the pupa matures into the

Tanzania armyworm outbreak. East Africa. Natives gathering army-worms for personal consumption.

armyworm moth, *Spodoptera exempta*. The "outbreak season," as it is called, usually occurs from mid-November through June or early July.

For pesticide spraying to be effective, early warning is necessary. Unfortunately, by the time the 1970 outbreak had been reported, the majority of the larvae had already become well-developed and had defoliated large portions of cultivated areas.

The East African Agriculture and Forest Research Organization has been conducting research to find an effective pesticide, but thus far, has not come up with a chemical that will control the armyworms.

83. AKITA-KOMAGATAKE VOLCANIC ERUPTION
Honshu, Japan   **September–December**

Akita-Komagatake Volcano began a new eruption on September 18, complete with ash columns, explosion sounds, incandescent cinders, and a small lava flow. The eruption was the first since 1932.

Emissions from the crater, 45 feet in diameter, included ash and lava blocks that shot 1,200

Akita-Komagatake volcanic eruption, Honshu, Japan. Night view of hauling cinders as seen from Komakusa-so on November 21, 1970.

feet into the air at five-minute intervals, and a 200-foot-wide lava flow that reached a distance about 1,800 feet from the crater. Eruptions with explosion sounds, cinders, and volcanic smoke took place 400 to 500 times a day in the first part of October. The activity became less dramatic in the latter part of the month, when the number of daily eruptions reduced to about 200. Volcanic tremors were also recorded by a seismograph installed at a hot-spring spa about two and a half miles northwest of the crater. In December, the volcanic explosions at the summit crater took place 100 to 200 times a day.

84. DWALENI METEORITE FALL
Nhlangano, Swaziland   October 12

Three meteorite fragments were recovered from an area near Nhlangano in the African nation of Swaziland. A high-pitched whine accompanied the objects' descent to earth, and explosions were reported as the meteor disintegrated. All three fragments were hard siderolites, similar in appearance, and distinctly magnetic. An 180-gram specimen analyzed at the Smithsonian Institution was found to be an olivine bronzite chondrite.

85. JAPANESE TANKER GASOLINE SPILL
Japan   October 16

An 800-ton tanker, the *Kasamatsu Maru,* exploded and sank on October 12 six miles off Cape Irozaki. Following the explosion, the cargo of 375,000 gallons of gasoline spilled in the sea. A neutralizing agent was immediately sprayed on the gasoline slick with considerable success.

86. ISLE OF WIGHT OIL SPILL
Isle of Wight, Great Britain   October 23

Two tankers, the *Pacific Glory* and the *Allegro,* crashed off the Isle of Wight on the evening of October 23. The *Allegro* made port safely, but the *Pacific Glory*, a 42,000-ton tanker with a cargo of 77,000 tons of crude oil, caught fire soon after the collision.

The ship grounded on a shingle bank four miles southeast of the Isle of Wight, and the fire was soon controlled. While almost 50,000 tons of the oil remained on board, a certain amount of fuel oil did leak out and floated ashore at Selsey and Brighton.

Although major ecological damage was averted by pumping the oil into other vessels, there was some limited pollution of adjacent coasts, especially to the east of the crash. Public and private organizations were fully mobilized to attempt to control the oil and watch over wildlife on the shore. According to the Royal Society for the Protection of Birds, only a few birds were washed up on shore.

87. KIFFA METEORITE FALL
Kiffa, Mauritania   October 23

Explosions, flashes of light, a smoke trail,

and a "chattering" sound accompanied the descent of a meteor over the desert town of Kiffa in the Sahara nation of Mauritania. An investigation conducted by the Management of Mines and Geology resulted in the discovery of many small meteoritic fragments weighing 1.5 kilogram in total.

The particles were found in a hole less than 20 centimeters deep in fine sand with no trace of burning. The specimens, therefore, had been relatively cool and brittle upon descent. The meteorite was estimated to have been less than 2 kilograms in total weight before shattering. The "chattering" noise heard by some people was probably the last fragments striking the vegetation in the impact area.

88. MADANG EARTHQUAKE
Madang, New Guinea   October 31

An earthquake with its epicenter located about 18 miles north of Madang caused hundreds of landslides, the deaths of eight people, and the destruction of hundreds of homes. The quake registered a Richter magnitude 7.0, but no evidence of faulting or sea level changes were reported.

89. ALCEDO VOLCANIC ACTIVITY
Galapagos Islands   October–November

Suspected volcanic activity at the Alcedo Volcano on Isabella Island turned out to be a large fire instead.

According to one observer, the fire in the crater was quite small in relation to the size of the entire rim, so wildlife was not affected. The volcano had the largest population of tortoises in the Galapagos Islands, but no dead or burned tortoises were reported seen. The cause of the fire remains unknown.

90. FORMENTERA WHITE-TAILED RAT INFESTATION
Balearic Islands, Spain October–December

Farmers in the Balearic Islands claimed to have been "invaded" by hordes of white-tailed rats, or, that is, a species of European dormouse. The farmers said the rodents had eaten their way through all types of crops, from fruits to grains. Studies showed, however, that the farmers may have exaggerated somewhat.

Although a high population of dormice is not unknown in the islands, this fall's crop of rodents was not really that unusual. Also, in-

vestigations showed that the dormice, which normally feed on insects, spiders, small lizards, and mice, had not eaten much vegetation. Tests of stomach contents showed very little plant remains and only a slight trace of chlorophyll.

No steps had been taken to exterminate the dormice, primarly because (1) there is no practical way of effectively exterminating them, (2) they apparently had not caused as much damage as the local farmers claimed, and (3) the extermination of the animals might seriously upset the ecological balance on the island.

91. MALAYSIAN FROG WAR
Sungei Siput, Malaysia   November 7–13

Ten or more species of frog, estimated as numbering more than 10,000, were reported as engaging in a fierce war. The "war zone" was in the vicinity of a Hindu temple and thousands of frogs fought in a brutal competition for several days beginning November 7.

The swamp was literally carpeted with battling frogs, ripping and tearing at one another, presumably warring over feeding and breeding grounds. The natives of the area, near the Malaysian seaport of Sungei Siput, claimed the war was an annual event.

However, zoologists from the University of Malaysia, after investigating the battlefield, discovered that the frogs were not making war at all. In fact, they were doing exactly the opposite: they were making love!

The scientists found the swamp waters filled with frog eggs and tadpoles. The so-called "war" was really a breeding aggregation or "mating orgy." The enthusiasm with which the frogs took to mating and the accompanying noise attracted a species of toads who joined in the celebration. Unfortunately, the toads gave off a highly toxic secretion fatal to the frogs.

92. BAY OF BENGAL TIDAL WAVES
East Pakistan (Bangladesh)   November 13

The most calamitous natural disaster of the century, a cyclone and accompanying tidal waves that roared up the Bay of Bengal into the Ganges and Brahmaputra river delta was responsible for the deaths of over 500,000 people. The death toll from the Bay of Bengal disaster was worse than the great China flood

of 1911, in which 100,000 were killed, or the Tokyo earthquake of 1923, responsible for 140,000 dead.

The storm surge destroyed one million acres of crops, 235,000 houses, and 265,000 head of cattle. The cyclone struck the coast of East Pakistan near the mouth of the Baringata River. The storm and tidal waves hit hardest at the hundreds of small fertile islands in the Bay of Bengal. The waves, whipped by winds more than 150 miles an hour, reached heights of 15 and 25 feet. Many small islands were completely submerged by the waves and remained under water for some time after the surge.

Two large islands, Bhola and Hatia, were nearly wiped out by the storm. Pilots flying over the area reported that half of Bhola had been totally destroyed, while on neighboring Hatia Island, the entire rice crop had been washed away. Villages disappeared into the storm and waves. A cargo ship weighing more than 150 tons was blown 50 yards inland on one of the islands. On 13 small islands near the port of Chittagong, not one inhabitant survived the storm.

The storm actually began on November 10 when a low-pressure area in the southeastern part of the Bay of Bengal concentrated into a depression located about 500 miles southeast of Madras and 800 miles southwest of Chittagong. The depression moved northwestward and intensified into a deep depression centered approximately 450 miles east of Madras; the low pressure mass moved into the west-central Bay and concentrated into a cyclonic storm with winds up to 54 miles per hour. The storm intensified as it approached the Bay of Bengal islands until winds reached 100 miles per hour. The shallow waters of the delta, pushed by the tremendous force of the storm, washed over the islands.

After the storm subsided, the battered people —some driven insane by the terror of the storm —wandered through the ruins searching for survivors and signs of life. Meanwhile, the government of Pakistan, located 2,000 miles away on the other side of India, simply ignored the plight of these suffering countrymen. The delayed rescue operations meant that thousands of the islanders died from starvation and disease.

The resentment created by governmental indifference eventually boiled into revolution and the formation of the new state of Bangladesh, independent of the Pakistan nation.

93. SCHUYLKILL OIL SPILL
Douglasville, Pennsylvania   November 13

An estimated three million gallons of "slop oil," or crankcase oil, was discharged into the Schuylkill River at Douglasville, Pa., when the earthen walls of a large man-made lagoon used by Berk's Associates Incorporated to contain the oil collapsed. The walls were apparently weakened by heavy rains.

A 50-mile stretch of the Schuylkill was affected, with the slick reaching the Delaware River as well. The shoreline was fouled by the slick and some 200 wild geese in Fairmont Park were contaminated. Damage to fish was minimal. However, all municipal water supplies on the two rivers were closed off.

94. SAN BERNARDINO BRUSH FIRES
California   November 13–17

Areas of San Bernardino and Los Angeles counties were devastated by innumerable brush fires that raged out of control for four days. Eighty-one square miles of brushland were destroyed, with the damage estimated at $16 million. No casualties resulted from the fires, although 52 homes, mostly in the resort community of Smiley Park, were destroyed. The destruction of the brush left the area susceptible to flooding and mudslides.

95. TYPHOON PATSY
Philippine Islands   November 20

At least 70 persons were killed and hundreds injured after Typhoon Patsy smashed through Manila and densely populated Luzon Island with winds that reached 124 miles an hour. The typhoon was considered the worst to hit the Philippines since 1882.

Tens of thousands of people in the Manila area were homeless as a result of the storm, and damage was estimated at hundreds of thousands of dollars. The area was declared a state of calamity by President Ferdinand E. Marcos.

The rain and winds buffeted Manila for about four hours, with gusts recorded up to 124 miles per hour. An estimated 25 percent of the city's homes were damaged. Three ships were grounded in Manila Bay and the city's domestic airport was shut down when two planes overturned on the runways. Most of the deaths resulted from flying debris, toppled homes or the washing away of homes by rampaging rivers.

96. MOUNT RIFFLER FIREBALL
Austria–Germany   November 24

A fireball of intense brightness entered the earth's atmosphere over central Europe and was photographically recorded by 11 German stations of the Max-Planck-Institut für Kernphysik and one Czech station at Ondrejov Observatory. The photographs were measured and computed at Ondrejov Observatory, but no meteoritic materials were found.

97. PANAMA POLLUTION POTENTIAL
Galeta Island, Panama   November 28

An oil barge owned by Refineria de Panama, carrying 5,000 barrels of bunker oil, washed ashore on the reef of Galeta Island, the site of the Atlantic Marine Lab of the Smithsonian's Tropical Research Institute. Efforts to free the barge from the reef were unsuccessful. After November 30 these efforts were discontinued, and the vessel was left as a constant threat to the marine life in the area.

Although the barge was not leaking any oil, the threat of possible spillage was well considered by Smithsonian scientists. In 1968, the staff fought to preserve their marine laboratory from oil contamination when the S.S. *Witwater* nearly fouled the island shoreline. The scientists considered two means for reducing the threat: one involves pumping the oil out of the barge or to a tanker offshore, or by pumping the oil across 100 yards of lagoon to the Refineria's trucks on shore. In any case, a pollution-control team was ready to contain any possible spillage.

98. LOUISIANA OIL PLATFORM FIRE
Gulf of Mexico   December 1

On December 1, a well operated by the Shell Oil Company ten miles off the coast of Louisiana and 65 miles south of New Orleans blew out of control, setting its platform on fire.

Some 12 of the platform's 22 wells caught fire. Ten were shut down quickly with automatic valves when the blast occurred. Each fire was systematically put out by drilling 11 relief wells and pumping water and mud into each of the burning wells. Attempts to clog the last well were unsuccessful until April 12,

1971, when it was finally blown out at the surface with pressure hoses.

Little oil spillage actually reached shore.

99. FLORIDA KEYS OIL SLICK
Florida   December 1

An oil slick reported by a fishing vessel stretched from Key Largo southward to Marathon Key near the Pennekamp Coral Reef. The slick was estimated as 75 miles long and a half-mile wide. Chemical analysis showed the substance to be diesel oil.

The slick was very close to the outer edges of the Pennekamp Coral Reef State Park. A fortunate change in the tides and winds, however, pushed the oil away from shore and the park. The slick dissipated after moving out to sea, leaving no effects on the wildlife in the park. The origin of the slick was not discovered.

100. JACKSONVILLE OIL SLICK
Jacksonville, Florida   December 1

Two Navy barges dumped their "sludge" oil, accumulated over a three-year period, 50 miles off the Florida coast in keeping with a 1926 law. The dump produced an oil slick 40 miles long and nearly 20 miles wide when the 500,000 to 700,000 gallons of oil were released into the water.

Ordinarily, the oil would have gone out to sea, but a shift of winds pushed the slick toward the shore. On December 3, two days after the slick had been discovered, two patches broke away from the main slick body. The first was observed about 17 miles east of Ponte Vedra (south of Jacksonville Beach); the other was seen 12 miles east of the shoreline between Ponte Vedra Beach and St. Augustine.

Florida officials and scientists believed that the slick was no threat to the wildlife on shore, but would adversely affect fish and other marine life if the oil sank.

101. CARIBBEAN MUD VOLCANO
ACTIVITY
Galerazamba, Colombia   December 2

On December 2, a half-mile-high column of fire and smoke erupted from the Caribbean Sea near Galerazamba, Colombia. The rare eruption at sea was probably due to methane gases released by a mud volcano located on a submerged sandbar. Mud volcanos are abun-

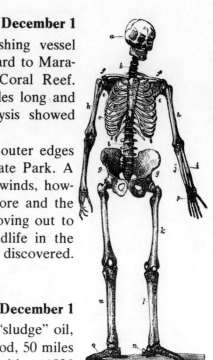

dant in the northern part of the country, and major eruptions of this type were observed in 1839, 1849, and 1958.

102. PERU–ECUADOR BORDER EARTHQUAKE
Peru–Ecuador   **December 9**

Approximately 30 people were killed and another 400 injured when a severe earthquake shook towns on the border of Peru and Ecuador. The quake registered 7.4 Richter magnitude and was located about 275 miles southwest of Quito, Ecuador.

The cities hardest hit by the tremors were Alamore, Selica, Macara, Cariamanga, Quiramba, Bellango, and Santanama, as well as several other smaller towns with populations between 10,000 and 20,000. With the exception of the Esmeraldas quake in 1958, the tremors were the strongest ever recorded in that area. After the main shock, more than 50 aftershocks of varying intensity were felt.

103. LITORAL DE PISCO FISH KILL
Pisco, Peru   **December 18**

Millions of fish washed ashore along the Peruvian coast at Pisco, forming a thick layer of dead fish 15 feet wide and nearly two miles long. The fish killed included flounders, rays, "corvinillas," "pintadillas," "ayanques," and "cabrillas."

The cause of the mortality was not known, although some experts believed it the result of water contamination by either immense colonies of dinoflagellates or toxic sewage.

104. FLORIDA APHID OUTBREAK
Southeast Florida   **December**

Despite the use of previously effective insecticides, potato plants were heavily infested by unusual numbers of green peach aphids, *Myzus persicae*. The outbreak, also involving pepper, tomato, and squash plants, was caused by a lack of rain and the reduction in numbers of the aphid's natural enemies including lady beetles, aphid lions, flies, and spiders.

Losses in some individual fields were estimated at one-third of the crop, while the loss to the overall southeastern Florida area was projected at one-fifth. Equally important, many aphids developed wings as the plants ripened and dispersed to other areas throughout southern Florida carrying virus infections fatal to non-potato plants.

105. NYRAGONGO VOLCANIC ACTIVITY
Zaire, Africa   **December**

The level of a lava lake in the interior of the Nyragongo crater has been reported rising at a rate of 120 feet a year since 1968. A red glow over the volcano had been observed several times during 1970, presumably the result of small lava flows filling up the lake. The temperature of the outflowing lava—from fissures in the northern and western part of an islet in the lava lake—was measured at 920 to 950 degrees centigrade. Throughout December, there were reports of weak seismic shocks, but their connection with the expansion of the lava lake could not be shown.

106. THE AMER RIVER OIL SPILL
The Netherlands   **December 27**

An electric power station oil supply tank located on the southern bank of the Amer River suddenly burst open on December 27, endangering a nearby complex of islands used as a winter wildlife refuge. After the tank ruptured, approximately half of its 16,000 tons of crude oil streamed over a low embankment into the Amer. Due to the low temperature, the oil was rather dense and floated on the river in a doughy mass some two feet thick. The mass, driven by an eastern wind, floated toward Brabantse Biesbosch, an inland delta area between the southern tributaries of the Rhine and Meuse rivers.

The spill resulted in the deaths of at least 5,000 birds, including the total mallard population of the Amer.

Most of the oil slick was soon removed with five draglines on barges. By January 6, Dutch authorities claimed that only 400 tons of crude oil remained in the Amer. Although other estimates showed that more oil remained, efforts to cleanse the river were officially curtailed until a contingent of 100 military men began moving oil by hand on January 25.

The mid-January thaw, however, aggravated the situation, and oil invaded the once ice-blocked creeks of the Biesbosch, severely contaminating stream bank vegetation. Although the government claimed that efforts would be made to remove the oil remnants, a government official announced on July 9 that the remnants could not be mechanically re-

The Amer River oil spill, the Netherlands. Reed vegetation on the oil-soaked Amer River.

moved without significantly disturbing the scientific value of the nature preserve. As a result, the oil was allowed to degrade on the spot.

The species of birds killed as a result of the spill included: tufted duck, wigeon, greylag goose, white-fronted goose, Bewick's swan, gulls, moorhen, and coot.

107. WEST MALAYSIA FLOODING
West Malaysia   **December 30**

A monsoon moving southward struck Malaysia for a week beginning on December 30, causing the overflow of the Kelantan, Pahang, Perak, Johore, Klang, and Gombak rivers and their tributaries. The flooding, worse than the Great Flood of 1926, resulted in the deaths of 39 persons.

The waters of the Kelantan, Pahang, and Johore rivers rose in some places 80 to 100 feet. Although the extent of the damage was not immediately known, the floods destroyed roads, railway lines, bridges, buildings, and kampongs. Approximately 50 major landslides were reported, 40 of them in Pahang State.

A state of national disaster was declared on January 5. Disease control was maintained by mass vaccinations against cholera and the evacuation of about 200,000 people from the flood areas. On January 6, the flood began to recede slowly in most areas, except in Pahang, where two attempts at blasting the sand bank at the mouth of the Pahang River did not help accelerate the flow of water out of the flooded region.

Guayas Province flooding, Ecuador. *Aerial photograph obtained through the courtesy of the United States Information Service Station of the Consulate General in Guayaquil, Ecuador.*

EVENT REPORTS   1971

Saint-Jean-Vianney landslip, Quebec, Canada. View from the southwest where a series of massive clay cave-ins swallowed up at least 40 homes. *Photo by R. M. Issacs. Courtesy of Geological Survey of Canada.*

1. NORTH ATLANTIC PUFFIN DECLINE
Northwest Atlantic Ocean   January

According to Dr. W. R. P. Bourne of the University of Aberdeen, the puffin, *Fratercula arctica,* the smallest and most pelagic auk breeding in the temperate North Atlantic, has been known to be declining in the southeast part of its range for at least half a century. Colonies formerly estimated at hundreds of thousands in Brittany, Cornwall, and the approaches to the Irish Sea are now numbered only in tens, hundreds, or at the most thousands. Also, a recent census suggested comparable, though less marked, declines in western Scotland, Norway, and a most important colony on St. Kilda, 45 miles west of the Outer Hebrides.

Dr. J. J. M. Flegg, director of the British Trust for Ornithology, reports that burrow counts and other studies during the three summers 1969–1971 indicated that estimates of a population of up to three million pairs up to the early 1960s did not seem unreasonable. Since then, however, there has been a continuing marked decline. The mile-long island of Dun was once completely riddled with occupied puffin burrows; in 1969 only about half the island was occupied, and in 1971 only about one-tenth remained occupied and free of an overgrowth of sorrel. It seems most unlikely that the present population of the whole island group exceeds 250,000 pairs.

No explanation can be found for the decline. No excessive mortality of puffins has been noticed in recent years nor do they figure largely in observed mortality due to oil pollution. Several researches even report that puffins contain smaller toxic chemical residues than other birds. There is also no evidence for recent epidemic disease, apart from a report that they have occasionally been found suffering from avian pox. The local herring fisheries are currently very productive, though those in the southern North Sea and off Scandinavia have failed recently.

A certain number of gulls prey on the puffins in their breeding quarters, but smaller puffin colonies are increasing in the presence of more gulls along the east coast of Britain. It should, however, be noted that the main French colony of puffins on the Sept-Îles decreased from 2,500 to 400 pairs after the *Torrey Canyon* disaster with little visible evidence of mortality due to oil.

2. SAKAKAH FIREBALL
Saudi Arabia   January 8

A Saudi Arabian radio operator in the town of Sakakah was aroused by a flash of light that brightened the evening sky for about a minute. The operator went outside and almost immediately heard and felt shockwaves severe enough to break windows in several buildings in the town.

A meteorite was determined to be the cause of the light and shocks. Although it was concluded that a meteorite might have been large enough to leave an impact crater visible from the air, an airborne search the next day was unsuccessful.

3. SAN CLEMENTE BEACHED WHALES
San Clemente Island, California   January 8

A pod of 28 dead pilot whales, *Globicephala scammomi,* were found stranded on the beach at Pyramid Cove, San Clemente Island. The animals, mature females and young whales of both sexes, were strewn along the high-tide line for about 200 yards.

San Clemente beached whales, San Clemente Island, California. Pacific pilot whales at Pyramid Cove on San Clemente Island January 10, 1971.
Photo courtesy W. E. Evans. Dept. of Navy. Naval Undersea Research & Development Center (NUC). San Diego. California.

The carcasses were too decomposed for extensive tests, but Navy scientists who conducted examinations concluded that the grounding was a natural and somewhat ordinary event comparable to similar strandings recorded in the region for the past 200 years. They also believed that meteorological and biological conditions—high tide, no surf, a very slight wind, a sloping beach with a steep drop-off and the presence of spawning squid—created optimal conditions for a whale stranding.

4. WEST IRIAN EARTHQUAKE
West Irian, New Guinea   **January 10**

The strongest earthquake reported by the Smithsonian Center since 1969 struck New Guinea approximately 80 miles southwest of Sukarnapura (Hollandia), registering Richter magnitude 8.1. Although there were only scattered reports of substantial damage, the quake was powerful enough to make inoperative the seismograph recording systems at Italy's Bendandi Observatory and the Weston Observatory in Massachusetts.

The quake was felt in most of northern West Irian. In Djajapura, cracks developed in brick walls and ten wooden structures built on pillars floating on water collapsed completely. In Sentani, 25 miles away from Djajapura, a church wall was cracked; and farther inland, about 14 wooden pillar-houses toppled. In Genjem, 25 miles away from Sentani, a sound was heard resembling gunfire, after which earthslumps and fissures were observed erupting mud and sand.

The quake was the strongest since August 11, 1969, when a shock of magnitude 8.1 struck in the southern Kurile Islands. In the New Guinea region, the last major earthquake (magnitude 7.2) occurred on October 31, 1970.

5. STRAIT OF DOVER OIL SPILL
Folkestone, England   **January 11**

Two tankers, the *Paracas* and the *Texaco Caribbean*, collided in the Strait of Dover. A day later, a third tanker, the *Brandenburg*, struck the wreckage and sank. Some 1,000 tons of fuel oil were released from the *Texaco Caribbean* and 500 tons from the *Brandenburg*. The oil was quickly brought ashore at Kent by a southeast wind. Although there were some seabird deaths reported, there was little other damage. The coast, which is mainly shingle or sand beaches and chalk cliffs, has few natural sites likely to suffer severely from oil contamination.

6. WURZBURG FIREBALL
West Germany   **January 17**

A brilliant fireball, with many sudden changes of brightness, including seven flares, entered the earth's atmosphere over West Germany and was photographed by six stations of the European All-Sky Camera Network. The photographic data indicated no significant mass reached the earth.

7. SAN FRANCISCO BAY OIL SPILL
California   **January 18**

Two Standard Oil of California tankers collided in heavy fog just outside of the Golden Gate Bridge, causing approximately 1.8 million gallons of bunker-type crude oil to wash ashore along San Francisco and Marin counties ocean coasts and a short distance inside the Bay. By January 21, globs of oil were being washed ashore adjacent to Point Reyes National Seashore and as far south as Pigeon Point, San Mateo County.

The California Department of Fish and Game office in San Francisco assumed control of processing oil-covered birds and began gathering them from major collecting centers. The San Francisco Zoo had prepared a bird care and cleaning center in the basement of their "lion house," and 1,679 birds were cared for at the zoo. Another bird care center was established in Richmond where approximately 1,000 birds were cared for during the after-spill period.

Approximately 5 percent of the 7,000 birds affected survived. Grebes and loons succumbed quickly to the effects of oil and stress, while ducks and gulls had a higher survival rate. Several cleaning solutions were used including soap and water, mineral oil, basic H, and polycomplex A-11. The birds received vitamin tablets, raw fish, and water at periodic intervals. The holding area was kept at a constant temperature both day and night with human disturbance kept to a minimum.

8. UTAH SHEEP KILL
Garrison, Utah   **January 20**

A toxic weed, *Halogeton glomeratus*, was found to be the cause of death for some 1,250 sheep near Garrison, Utah. Although initial news reports implied that the deaths might have been due to either escaped nerve gas from the nearby Dugway Proving Ground or radiation fallout from Nevada atomic test sites, subsequent field and laboratory tests proved that Halogeton poisoning was the cause.

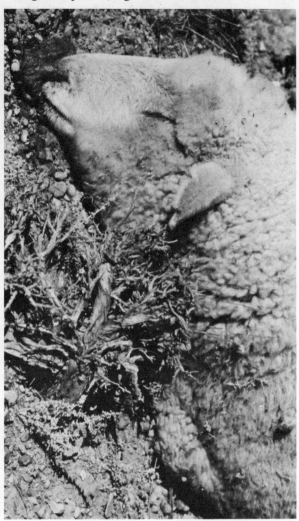

Garrison, Utah sheep kill. One of the 1,250 victims of the toxic weed Halogeton. *Photo courtesy Southwestern Radiological Health Laboratory, Las Vegas, Nevada.*

Halogeton is a low, red, prickly plant with a high water content, making it particularly attractive to sheep. Although highly toxic to sheep, cattle can eat the weed without harm. The plant flourished on the affected Utah range, making up some 60 percent of the vegetation. Another 1,350 sheep present on the range apparently did not graze on the weed, and they were not affected.

Periodic outbreaks of Halogeton poisoning have been reported in the past, and a major incident in Idaho (Raft River) in 1948 killed some 1,000 sheep.

9. LONG ISLAND SOUND OIL SPILL
Connecticut   **January 24**

A 685-foot tanker, the *Esso-Gettysburg*,

owned by Humble Oil and Refining Company, ran aground on a rock ledge at the mouth of New Haven harbor sometime before dawn. A series of holes torn in the port side of the ship leaked an estimated 363,000 gallons of oil into the harbor and Long Island Sound.

The oil slick that resulted spread rapidly out on the water of the outer harbor and the main slick eventually covered an area approximately 10 miles long, extending from New Haven to Guilford. The incoming tide and strong southerly winds carried oil to upper shore beaches. During the following days prevailing westerly winds, often of gale force, carried much of the oil in an easterly direction with fingers of oil extending as far as Hammonasset Beach, Madison.

The Board of Fisheries and Game first became aware of the problem late January 23. Investigations were initiated that afternoon and the following morning to determine the extent of damage to wildlife.

A shoreline census of waterfowl from the Branford River to Sachem Head, Guilford, disclosed approximately 600 scaup, 35 miscellaneous divers, 118 mallards, and eight black ducks within approximately one mile of the shore. These birds, totaling an estimated 761, appeared to be in the close proximity of the contaminated area and were considered vulnerable to the oil.

A census of dead ducks, started on January 23, reported that 63 dead ducks had been counted in the area from West Haven to Guilford. Of the dead ducks counted, approximately 17 percent were dabbling ducks and 83 percent were diving species.

With the exception of two mallards, all the dabblers seen were black ducks; the dead diving ducks counted included 44 scaup, four golden eyes, three canvasbacks, and one bufflehead. In addition to ducks other dead species counted included three herring gulls, one horned grebe, and one cormorant.

The fact that only a day or two previous to the spill many of the scaup in the spill area had moved to the west probably reduced the numbers of affected waterfowl.

10. CERRO NEGRO VOLCANO
Nicaragua   **February 3**

Cerro Negro erupted again on February 3, 1971, thus marking its fourteenth month of eruptive activity.

Cerro Negro Volcano, Nicaragua. Electrical phenomena above summit crater. *Photo courtesy Ing. F. Peñalba. Estudios Peñalba. Nicaragua.*

The eruption, similar to the 1968 activity, was characterized by the continuous violent eruption of gases, bombs (pump-like ejections), stone blocks, lapilli, and ash. A cloud reached an altitude of 20,000 feet before being dispersed by prevailing winds. The explosions of the first day of activity came approximately every 10 seconds, and formed continual turbulent clouds.

Between the cloud and the crater, or main cone, continual electric phenomena were observed, possibly caused by static charges formed by particle friction and by the great temperature differences existing inside the cloud. The bombs, scoria, and stone blocks were thrown up to 500 meters from the crater.

On February 9, the eruption changed character, passing from the continual eruptive phase to an intermittent phase. The cycles in the first days were 15 minutes of activity and 10 minutes of calm; as the days passed, the periods of calm became longer until activity ceased completely on February 14.

The 1971 eruption of Cerro Negro was one of the most violent in recent times, possibly because no openings of adventitious craters were produced in the base or flank of the main cone; energy was therefore retained totally in the main crater. The eruption of 1971 occurred exclusively from this crater and from there it was particularly violent. However, during the sixth day, a small amount of activity was observed in the adventitious crater of 1957 which is situated to the east of the main crater and almost totally covered by pyroclastic materials.

The changes produced during the 1971 eruption included a great widening of the main crater, which grew from 450 feet at the greatest diameter to 1,200 feet; widening of the diameter of the base of the main cone, which grew from 1,850 to 3,300 feet; a general increase in height; and a partial or total concealment of numerous old vents from 1923, 1947, 1957, 1960, and 1968.

11. PRINCETON FIREBALL
East-Central U.S.A.   **February 3**

A fireball—as bright or brighter than the full moon—was observed over hundreds of square miles as it traveled southwesterly between Dayton, Ohio, and Memphis, Tenn. The meteor was reported as bright green or blue-green with a "tail" seven times longer than the diameter of the fireball. Most observers reported seeing it for no more than two to four seconds. An observer located six miles south of the town of Princeton, Ky., in the far western part of that state, reported a sonic boom.

12. KEPULAUAN MENTAWI EARTHQUAKE
Sumatra, Indonesia   **February 4**

An earthquake reported by the National Earthquake Information Center at Richter magnitude 7.1 struck the Kepulauan Mentawai Islands on February 4. The Soviet Seismic System also notified the Center and placed the magnitude at 7.5 Richter.

13. TUSCANIA EARTHQUAKE
Tuscania, Italy   **February 6**

The ancient town of Tuscania, located some 50 miles north of Rome, was practically obliterated on February 6 when twin earthquakes crushed the city within minutes. The quakes struck the town and surrounding villages, killing at least 18 persons, injuring another 270, and leaving the town's 7,000 inhabitants homeless. An aftershock struck the next day, destroying some of the few buildings left standing. The damage was estimated at $41.6 million; it was the worst disaster in Italy since a 1968 earthquake in western Sicily killed 316 persons and left another 90,000 without homes.

Inhabited for some 23 centuries, Tuscania was the home of rare and valuable Etruscan art, much of it preserved in pre-Christian tombs. The damage to these artifacts was considered as serious as the damage to the art treasures in Florence during the 1966 floods there. Ironically, the quake was not exceptionally strong, measuring only 5.0 Richter magnitude. Nevertheless, the shocks were violent enough to destroy 100 buildings in Tuscania and damage 400 in the town's medieval center, and strong enough to be felt as far away as Rome and Trieste. The nearly total destruction and the danger to surviving buildings forced residents to evacuate their homes.

14. ANTARCTICA EARTHQUAKE
Antarctica   **February 8**

An earthquake, 7.0 Richter magnitude or higher, shook the Argentine Antarctic territory, but did no damage to the scientific research bases on that continent. Reports from the Argentine Antarctic Institute stated that the tremor was followed by sea swells and ice

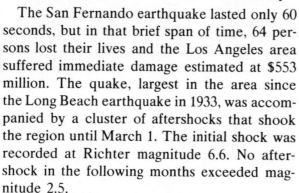

cracks, but that the penguin colonies and the rest of the plant and animal life in the region were not harmed.

15. SAN FERNANDO EARTHQUAKE
California **February 9**

The San Fernando earthquake lasted only 60 seconds, but in that brief span of time, 64 persons lost their lives and the Los Angeles area suffered immediate damage estimated at $553 million. The quake, largest in the area since the Long Beach earthquake in 1933, was accompanied by a cluster of aftershocks that shook the region until March 1. The initial shock was recorded at Richter magnitude 6.6. No aftershock in the following months exceeded magnitude 2.5.

After an analysis of all damage, including factors such as loss of tax revenue, the eventual economic impact on the Los Angeles area was projected as $750 million. A total of 180 schools were damaged. Of these, 18 had to be vacated, and major structural damage was reported in another 35.

damaged, and of these, 465 were declared unsafe. In addition, there were 62 apartment houses and 372 commercial structures so severely damaged that they were also declared unsafe.

Destruction to highways and roads was estimated at $27.5 million—$22.5 million to the state freeway system and $5 million to city and county roads. The major freeway damage resulted from the collapse of five overpasses and partial obstruction of an additional seven. Five dams were damaged—the Upper and Lower Van Norman dams, Pacoima Dam, Lopez Dam, and Hansen Dam. The Lower Van Norman Dam was damaged so severely that the estimated cost of replacement is $34 million. The gravity-arch Pacoima Dam suffered about $1.5 million damage at its abutments, while minor damage was reported at Lopez and Hansen dams.

Public structures damaged—exclusive of dams, highways, and hospitals—included those owned by the city of Los Angeles, the city of San Fernando, the county of Los Angeles, the state of California, and the federal government.

Karua submarine volcanic eruption, New Hebrides Islands, South Pacific. *Photo courtesy of National Geographic Society.*

Two hospitals, the recently built $34-million Olive View and the 50-year-old $15-million Veteran's Hospital, sustained major damages. The partial collapse of both structures probably will lead to their replacement. The Veterans' Hospital was constructed in 1920, before the establishment of earthquake codes in California.

A total of 21,761 single-family dwellings were

Damage to structures owned by the city of Los Angeles alone was estimated at $73 million.

16. KARUA SUBMARINE VOLCANIC ERUPTION
New Hebrides, South Pacific **February 22**

The Karua submarine volcano erupted in a

monumental display on February 22, hurtling clouds of steam and black masses of cinder, ash, and presolidified crust up to 600 feet in the air. The activity was accompanied by frequent explosions and thunder and lightning precipitated by the intense heat of the eruption.

The eruption suddenly ended 10 hours after it had begun; but not before creating a low, flat island over a half-mile wide with large boulders strewn about its surface.

Ten days after the eruption, a New Hebrides scientist was able to walk on the newly formed land mass created by the activity, an islet that he likened to "moonlike terrain." The ground was still warm and a distinct smell of sulphur was in the air. On one side of the islet, the sea was still too hot to touch. Investigators assumed the actual crater and the nucleus of volcanic activity was located at this spot.

17. LUDWIGSHAFEN FIREBALL
West Germany   February 23

A widely observed fireball entered the earth's atmosphere over West Germany near Ludwigshafen on the Rhine River. Although picked up as an unidentified object by an American radar station near Mannheim, the object was not photographed by the western sector of the European All-Sky Camera Network, which went into operation only seven minutes after the event. The eastern part of the network was not able to photograph the object because of a cloudy sky.

Enough visual observations were made to warrant an intensive search by the Max-Planck-Institut für Kernphysik. However, nothing was found resembling a meteorite or satellite fragment.

A fireball, seen over northern Italy and southern France 30 minutes after the West Germany event, was thought to have been the same object. Later, the French newspaper *La Marseilleise* reported that the "fireball" was not extraterrestrial material, but merely the result of rocket experiments conducted in France.

18. *WAFRA* OIL SPILL
Cape Agulhas, South Africa   February 27

The tanker *Wafra*, carrying 63,174 tons of crude oil, sprang a leak and was grounded on a reef off Cape Agulhas, causing the beach area between the Cape and Quoin Point to be contaminated. During salvage efforts more

tanks were damaged and eventually nine of the *Wafra*'s 18 tanks were leaking. The resulting oil slick measured 20 miles long and some three miles wide.

Within a matter of days, a fleet of ships both sprayed and dumped enormous amounts of relatively nontoxic dispersing chemicals to rid the deep water of the oil slick. No chemicals, however, were used in the intertidal zone. The beach-cleaning operation was severely hampered by the rocky and often inaccessible nature of the coastline. A remarkable degree of success was achieved by applying straw to the oil slick. The straw absorbed the substance and was subsequently collected and removed to a dumping site.

An effort to tow the *Wafra* failed because of inclement weather, and the ship was sunk using missiles and depth charges. During the towing and sinking, however, more oil was leaked into the sea. Various surveys showed that damage to wildlife was slight. The animals hardest hit by the oil leak were giant periwinkles, sea urchins, and limpets. Over 1,000 penguins from a population of some 15,000 birds on Dyer Island were treated for oil contamination. Seaweed and other marine plants, as well as most fish, were not seriously affected by the spill.

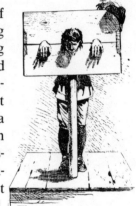

19. ERTA 'ALE VOLCANIC ACTIVITY
Ethiopia   March

The activity in both lava lakes of the Erta 'Ale Volcano became almost continuous after March, although no extra-crateric eruptions occurred. The volcano is known for its permanent lava lake, a rare form of volcanic activity, found now only in Africa's Nyragongo Volcano and, before 1924, in Hawaii's Kilauea Volcano.

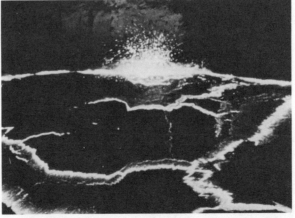

Erta Ale volcanic activity, Ethiopia. Night view of the Erta 'Ale central pit crater, lava lake activity, and fountaining on February 27, 1971. *Photo courtesy of J. Varet, Haile Selassie I University, Ethiopia.*

Reports throughout the year stated that the main surface of the northern lava lake formed in December 1970 was unchanged, and that its solidification was still continuing. The northern lava pool, however, was reported still very active. White smoke is always present over the volcano: hence its name, which means "smoky mountain" in the Afar language of Africa. Since 1906, when the phenomenon was first reported, the presence of red lava has been observed.

20. *THUNTANK 6* OIL SPILL
Milford Haven, South Wales   March 19

The Swedish tanker *Thuntank 6* ran aground on Thorn Island with damage to the hull resulting in the loss of 159 tons of light fuel oil. Most of the oil was successfully dispersed at sea in the western end of Milford Haven using the emulsifier BP 1100. Some oil, however, came ashore at several sites on both the north and south shores of the Haven.

Cleaning operations were carried out under the supervision of the Gulf Oil Refining Company using BP 1100 and suction pumping equipment. Although no oiled birds of any species were reported following the spill, a significant mortality of the limpet, *Patella vulgata,* was reported at West Angle Bay.

Sandy Haven was visited one week after the spill, when only very slight traces of oil were apparent. Large numbers of cockles, *Cardium edule,* were found in the unusual position of being completely above the surface of the sand, some obviously dead, others gapping very slightly. On returning 200 of these to a laboratory for recovery, a 99 percent mortality ensued within five days, compared with a 3 percent mortality in the control group of cockles collected from an unpolluted shore. No traces of oil were found either on the cockles or in the surrounding sediments, indicating the high cockle mortalities to be due to some other factor, possibly the toxic effect of the emulsifier.

In general, however, it appeared that very little biological damage resulted from the *Thuntank 6* oil spill, either on a short- or long-term basis. The reasons for this include the efficiency of dispersing operations while most of the oil was still at sea, and the careful use of low-toxicity emulsifiers on shore. The oil itself appeared to be of a relatively non-persistent type and of very low toxicity to shore life.

21. CHUNGAR AVALANCHE
Chungar, Peru   March 18

The copper and lead mining community of Chungar, some 90 miles northeast of Lima, was completely destroyed when a mass of rock slid into Lake Yanahuani creating a fatal water wave 20 yards high. Approximately 200 persons living in the small town were killed in the flood.

Although the disastrous landslide occurred on the same day as a fairly strong earthquake, the quake was not thought to be the direct cause of the avalanche. The rocks of the mountain, cracked previously by weathering, had begun to move because of a heavy rainfall. The rocks reportedly slid down the mountainside into the lake at speeds estimated as high as 150 miles per hour.

22. SOVIET PIPELINE LEAK
Ural River Valley, U.S.S.R.   March 21

A break in the 800-mile pipeline on the west bank of the Ural River between the cities of Guryev and Uralsk threatened fertile bottom lands of nearby collective farms and the sturgeon fishing grounds of the Caspian Sea. The Soviet government provided no immediate information on the damage caused by the leak; however, the government newspaper, *Pravda,* did state that the delay of control measures would result in serious damage to the area.

According to Theodore Shabad, Moscow correspondent for *The New York Times*, maintenance crews were able to contain the spilled oil behind temporary banks of earth thrown up by bulldozers although it was feared that the embankments of frozen ground might give way in the spring. The *Times* also reported that oil was already beginning to seep through the banks and that water from the melting snow was expected to carry the oil into the Ural.

Because the crude oil has an unusually high wax content, heating units had to be installed at various intervals of the line to keep the oil fluid and moving. According to *Pravda,* the pipeline was unable to withstand the strain produced by the differences between the low outside temperature and the relatively high temperature of the oil inside the line.

23. NYAMURAGIRA VOLCANIC ACTIVITY
Zaire, Africa   March 24

The Nyamuragira Volcano erupted on March

24, emitting a massive lava flow and volcanic cinders that caused serious damage to crops and livestock within a 30-mile radius. Located in the Albert National Park, the volcano contains a number of adventive cones on its flanks and in the surrounding area. By May 2, the more spectacular activity had almost totally ceased.

24. *TEXACO-OKLAHOMA* OIL SLICK
Cape Hatteras, North Carolina   March 27

A Boston-bound tanker, the *Texaco-Oklahoma*, broke in two and sank 120 miles at sea on the morning of March 27. The ship, carrying 220,000 barrels of heavy-sulphur fuel oil, had apparently been caught in a gale and high winds in an area off Cape Hatteras often called the "graveyard of ships."

An oil slick extending some 50 miles around the sinking site did no harm to wildlife.

1971

Although there were no reported leaks on the *Panther* and no oil spill was seen during the transfering operations, the oil that came ashore was analysed and found to be similar to the Venezuelan crude oil aboard the ship. Still, it was impossible to prove conclusively that the oil actually came from the tanker *Panther*.

On April 2, the oil was reported ashore at Deal, and adjacent beaches from Kingsdown to Sandwich were contaminated to a moderate degree. Owing to offshore winds, only limited pollution occurred in Britain. Only one or two dead birds were found, although 200 gulls and ten fulmars were reported oiled. The bulk of the oil was dispersed by winds into the English Channel.

Fortunately, since it was a mild spring, the large flocks of auks and sea ducks that winter in the area and later pass through the Strait of Dover on migration had already moved north before the pollution.

Ruapehu volcanic activity, North Island, New Zealand. *Photo courtesy Dr. J. H. Latter, D.S.I.R., Wellington, New Zealand.*

25. HAWAII FIREBALL
Hawaii   March 29

A fireball trailing a cloud of tiny sparks and producing five small, distinct objects was seen traveling in a northwestern direction over Hawaii. The largest object glowed with an intense blue-white light, followed by the four smaller, red-orange objects.

26. ST. MARGARET'S BAY OIL SPILL
Dover, England   March 30

The Panamanian-owned tanker *Panther*, carrying 25,000 tons of Venezuelan crude oil, grounded off the coast of Dover. Although the tanker was eventually towed from the site and part of its cargo was removed, an oil slick remained to threaten the English coastline.

27. RUAPEHU VOLCANIC ACTIVITY
North Island, New Zealand   March–May

Minor steam and mud emissions, plus variable volcanic tremors, characterized Ruapehu activity from March to April. On May 8, however, the volcano erupted in a series of spectacular explosions. The explosions uplifted mostly crater lake water, but there were also considerable amounts of earth blocks, lake sediment, ash, and mud ejected to heights of nearly 500 feet.

28. STROMBOLI VOLCANIC ACTIVITY
Kipari Island, Italy   April 3

The Stromboli Volcano erupted in a lava flow on April 3, with the flow occurring on the

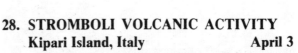

97

active Sciara del Fuoco, between the crater and the sea on the northwest flank. On May 1, two explosions reopened a crater blocked since the March 31 explosions, and a small amount of lava was emitted. By May 4, after days of small vapor emissions, the activity had returned to its normal state.

29. MOUNT ETNA VOLCANIC ACTIVITY
Sicily, Italy   April 5

From January 1966, Mt. Etna had been almost continuously in turmoil, with activity characterized by both explosions within summit craters and by slow emissions of lava. This persistent, but generally moderate, activity culminated in the massive outbreak of 1971.

The eruption of Mt. Etna began on April 5 with the opening of two radial fractures about 600 feet apart, at the foot of the central crater. Lava flows poured out at the lower part with initial speeds up to 10 miles per hour. During April, lavas flowed on the south flank of the mountain, destroying the observatory and the upper part of the cableway there.

Mount Etna volcanic activity. Sicily, Italy. *Photo courtesy of J. C. Tanguy, Lab. de Géomagnétisme, Paris, France.*

A new activity center in the lower regions of the volcano produced huge quantities of lava at a more or less constant rate for three weeks. The erupting magma was highly fluid and foamy. It invaded the pineland of Cubania and then spread to the rich lower eastern slope of Mt. Etna, seriously threatening the villages of Formazzo and Saint 'Alfio. The lava stopped in the valley of Cava Grande, after flowing for approximately five miles.

On May 18, an explosive outbreak began from a circular, newly opened abyss at the upper end of the fracture at the 6,000-foot level. For many days, high pressure gases, throwing cinders and blocks into the air, continued to escape from the crater. Although it was difficult to approach the most active lava flows, temperatures as high as 1,130 degrees centigrade were recorded near the various vents. During the first days of June, the discharge of lava began to decrease, and on June 12, the eruption was over.

30. WETHERSFIELD METEORITE FALL
Connecticut   April 8

Since 1847, only 11 meteorites have been known to strike buildings. The 11th crashed into the roof of the home of Mr. and Mrs. Paul Cassarino in Wethersfield, Conn., a suburb of Hartford. The fireball that produced the meteorite was spotted just before dawn by a West Hartford resident and described as a "bright streak like a bomb bursting in the east-south-east," traveling roughly across Hartford in the direction of Wethersfield.

Mr. and Mrs. Cassarino apparently were asleep when the 12.3-ounce meteorite hit. At 6 a.m., when Mr. Cassarino awoke for work, he discovered plaster on his living-room floor. Looking up, he saw the meteorite embedded in the fiberglass insulation of the ceiling, and by standing on a chair, he was able to remove it.

The sample was sent to the Smithsonian Astrophysical Observatory and the Smithsonian Institution for analysis.

31. EAST COAST STARFISH KILL
North Carolina–Maryland   April 16

Thousands of dead starfish were reported washing up on the Atlantic coast from Cape Hatteras, N.C., north to Rehoboth Beach, Md. Dead starfish were first seen at Nags Head Beach just above Hatteras; and again, the fol-

Mount Etna volcanic activity, Sicily, Italy. Eastern fracture and lava fountains, April 7, 1971. *Photo courtesy J. C. Tanguy, Lab. de Géomagnétisme, Paris, France.*

lowing weekend, when the Hatteras beach was reportedly covered with the dead animals. No explanation was given for the phenomenon. The starfish were either *Asterias forbesii* or *Asterias vulgaris.*

32. *NOCTILUCA MILIARIS* PLANKTON BLOOM
Spain   **April 17**

An unusually early appearance of the *Noctiluca miliaris* plankton along the Tarragona shore on Spain's Mediterranean coastline colored the waters there in great patches, some of them several hundred yards long. Although common in the northern part of the country along the coast of the Bay of Biscay, the phenomenon is very rare along this stretch of coast.

33. GWARZO METEORITE FALL
Ibadan, Nigeria   **April 25**

A shower of meteoritic material fell over farmland in Nigeria after, as one observer put it, "thunder-like sounds were heard and the descent of a reddish object with trails of cloudy smoke was seen." Four specimens were collected—with one stone, recovered in two pieces, weighing some 3 kilograms. The samples were olivine bronzite chondrites.

34. ANACORTES OIL SPILL
Anacortes, Washington   **April 26**

Approximately 55 barrels—or 230,000 gallons—of diesel oil spilled into Padilla Bay during a loading operation at the Texaco Refinery dock at Anacortes. During the day, swift tidal currents carried most of the oil out through Guemes Channel, into Rosario Strait, and throughout a large area of Puget Sound.

Considerable biological damage was caused by the oil, including the contamination of clams, worms, shore crabs, and other shellfish. About 1,000 birds—loons, scarp, scoter, and horned grebe—were killed as a result of the spill.

An air survey reported oil sightings on some of the islands in the strait, with the heaviest concentration near the south end of Guemes Island where the slick was washing up on the beaches. Clean-up operations were begun during the afternoon, using straw as an absorbent. By April 28, there were no large concentrations of oil visible in the water. Clean-up action was completed on April 30.

35. SULE SKERRY ISLET BIRD KILL
Sule Skerry, Scotland   **April 26**

The lightkeeper on Sule Skerry, a small isolated islet and the home of some 60,000 pairs of puffins *(Fratercula arctica),* reported that about 60 of the birds had become contaminated with oil and many had died. The report was given some importance because puffins have been greatly affected by oil pollution in recent years. The pollution was suspected to have come from North Sea drilling rigs or tankers washing out their holds in waters off northern Scotland. No investigation was made due to the inaccessibility of the island.

36. SALVADOR FLOODS
Salvador, Brazil   **April 26–30**

The heaviest rains recorded since 1903 when the meteorological service was established in Salvador fell on that city April 26 through 30 causing considerable loss of life and property.

During one 24-hour period, over 15 inches of rain fell; more than 21 inches fell in 48 hours, and the total rainfall for the five days was more than 23 inches.

Salvador's annual rainfall averages about 80 inches; thus, more than one-fourth of the normal yearly total fell in two days.

The rains caused several hillsides covered with houses to collapse, burying still other dwellings below. Streams became torrents and swept over their banks, destroying the houses clustered among them; valleys became lakes littered with floating household possessions. More than 100 deaths and 2,000 injuries were ascribed directly to the landslides and floods. Approximately 11,000 people were left homeless, and crowded into the houses of friends and relatives or into schools, markets, and warehouses converted into shelters by the state and municipal governments. Local authorities estimated that it will cost $5,232,000 to repair the many roads, bridges, and tunnels which suffered extensive damage.

37. AZUFRAL DE TUQUERRES TREMORS
Nariño, Colombia   **April 26**

Beginning on April 26, more than 60 local tremors were felt by the inhabitants of the village of Santander near the volcano Azufral de Tuquerres in Nariño, Colombia. The origin of the earthquake swarm seemed to be underground movement of volcanic origin without any external signs of volcanic activity.

Azufral de Tuquerres, a caldera with lake and lava domes, has never erupted, although fumarolic and solfataric activity has been recorded for centuries.

38. HONDURAS FOREST FIRES AND SMOKE POLLUTION
Honduras   **April 28**

Brushfires in numerous areas throughout the western two-thirds of Honduras created massive clouds of smoke covering the western third of the Gulf of Mexico.

The fires are normal during the dry season in Honduras because of the traditional slash-and-burn agricultural methods in which farmers burn off areas of underbrush in forest groves to clear pasturage for cattle.

In 1971, the fires were more widespread and

smoke pollution more serious than in other years. In many areas, planned burning went out of control and set large forest areas ablaze.

In late April and early May, visibility was poor over most of the Central American country. Prevailing northeast winds carried smoke from eastern Honduras toward the western part of the country, reducing visibility over the capital region and creating a serious problem for aviation. The onset of daily rains beginning around mid-May finally eliminated the smoke problem.

39. EVERGLADES DROUGHT
South Florida   **April–May**

Historically, the Everglades have always had cyclic periods of drought with accompanying brushfires. In fact, the biggest problem of land management in the Everglades is fire prevention.

The 1971 drought was most critical in mid-May, just before the rains finally came. At that time, several thousand acreas scattered widely over the Everglades were affected by fires. The damage from these fires would not have been so extensive, some experts think, if normal control burns had been conducted during the previous winter or if the general water level had been higher.

Despite the widespread burning, there was no major or permanent damage to wildlife. Although many of the fresh-water fish were killed when ponds dried up, the alligators, birds, and small animals found some relief in remaining potholes.

While the fires may be detrimental to some small game animals, the general population turnover of all animal life more than offsets any damage that may have occurred. Indeed, the fires are often beneficial to the large animal life, such as the black-tailed deer.

Regardless of water level, natural fish kills often occur in the extreme south of Florida during the summertime, primarily due to oxygen deficiency. However, the recovery rate of the warm fresh-water fisheries is so rapid that it is really of little or no consequence whether an entire area is killed. Generally, as soon as the water level rises, the fish population returns. Some of the main species of game fish found in Everglades fresh-water ponds are bass, blue gill, and pickerel.

40. GALAPAGOS EARTHQUAKE SWARM
Galapagos Islands   **April–June**

In early April, the National Ocean Survey reported three earthquakes within six miles of Isla Fernandina, the site of a massive caldera collapse in 1968. The earthquake swarm continued into early June, but inspection of the Fernandina Volcano, where the tremors were centered, showed no resumption of the caldera collapse begun in 1968. The Fernandina caldera had, in fact, remained unchanged as had other calderas on nearby islands.

41. SAIGA ANTELOPE INVASION
Northwest Kazakhstan, U.S.S.R.   **May**

A rare animal species, the Saiga antelope, has been so well protected by Soviet game laws that its growing herds were invading wheat lands, causing wide damage.

A wildlife official warned that miles of expensive fencing might have to be installed to keep the rampaging antelope under control. He also asked that hunting restrictions be relaxed so that more of the animals could be killed for meat each year.

D. Boiko, chairman of the Republic's Hunters Alliance, reported that the Saiga, once believed near extinction, now numbered one million head—double the population of five years ago. He predicted an increase to two million in coming years. According to the Kazakhstan hunting official, the largest herds, 800,000 head, spend the winter in the remote desert country of Bet Pak Dala, in the southern part of the republic. The animals die in large numbers in heavy snow and cold winters. In the spring, the herds have been migrating northward to summer pastures to escape the desert's heat and aridity. In recent years, however, the Saiga have migrated to newly developed wheat lands in the Turgai region of northwestern Kazakhstan.

Attempts had been made to chase the herds by helicopter, but they returned as soon as the aircraft disappeared. Boiko predicted that as many as 100,000 might feast on wheat shoots this year and warned that the herds might spread disease to livestock in the virgin lands. The official disclosed that the Kazakhstan Ministry of Agriculture was instructed in 1966 to set aside permanent pasture lands along the migration routes; but that the plan was never carried out, causing the invasion of the wheat areas.

42. *SQUILLA ARMATA* POPULATION INCREASE
South Africa   **May–June**

In May and June, large numbers of *Squilla armata*, a small crustacean, were found in the stomachs of snoek *(Thyrsites atun)* caught and examined on research boats from the Division of Sea Fisheries, Capetown. There had been no previous record of such intensive feeding on *Squilla armata* by snoek.

Also there were reports of large numbers of the crustacean close to shore, which was considered quite extraordinary. The scientists assumed that there either had been an unusual increase in population or a migration into areas not normally occuped by *Squilla armata*.

43. ADAK ISLAND EARTHQUAKE
Aleutian Islands   **May 2**

A strong earthquake shook both Adak and Amchitka islands, but no damage was reported as a result of the tremor. The quake's magnitude was recorded at 7.1 Richter. A tsunami warning was issued, but canceled less than an hour later.

44. SAN CRISTOBAL VOLCANIC ACTIVITY
Nicaragua   **May 3–early 1972**

San Cristobal, dormant for nearly 300 years, began activity on May 3 when its summit crater emitted gas visible from Leon and other nearby cities.

From May to early June, there were reports of small explosions and minor ash or sand fallout. Sulphurous gas was frequently reported, and trees within the crater on the downwind flank appeared to have died.

45. SAINT-JEAN-VIANNEY LANDSLIP
Quebec, Canada   **May 4**

A major landslide occurred on the west shore of the Petit Bras River at Saint-Jean-Vianney, Quebec. As a result, 31 lives were lost, and 40 single-family houses were destroyed and several others were severely damaged.

An estimated nine million cubic yards of soil moved in the slide from an area covering approximately 350,000 square yards. After complete liquefaction, most of the slide material traveled down the Petit Bras and the Rivière aux Vases at a speed of approximately 16 miles per hour for a distance of about 1.8 miles, and

into the middle of the Saguenay River. The reinforced concrete superstructure and the central pier of a 140-foot-long bridge over the Rivière aux Vases were dislodged and carried 500 feet into the Saguenay.

The Saint-Jean-Vianney region was covered about 9,000 years ago by an extension of the Champlain Sea in which thick layers of sensitive marine clay were deposited. This clay tends to liquefy under certain conditions of stress and strain, and landslides which develop within it are generally of the flow type.

of oil also permeated the river marshes and a rainbow sheen could be seen on the water.

American Oil began pumping out the line immediately after the break, but the oil dripped through the night at a rate of approximately a half-gallon per minute. On the morning of May 6, the line was completely shut off.

Although there were no reports of oil-coated birds, the oil was thought to have damaged some of the bottom-dwelling organisms and marsh vegetation. The clean-up operation was assumed by American Oil, who hired the In-

Saint-Jean-Vianney landslip, Quebec, Canada. *Photo by R. M. Issacs. Courtesy of Geological Survey of Canada.*

The slide occurred at a period of spring runoff after a winter with unusually heavy snowfall and a long, late spring. Between April 21 and 24, the rainfall was 0.71 inches, and on May 3 and 4, 0.73 inches fell.

Ten days prior to the disastrous May 4 slide, a landslide occurred on the west shore of the Petit Bras. It has been estimated from photographs to have been about 200 feet wide and 500 feet long.

46. YORK RIVER OIL SPILL
Yorktown, Virginia   **May 5**

A break in a pipeline leaked about 70,000 gallons of oil into the mouth of the York River, creating a film of oil two miles long, 100 yards wide, and 1/32 of an inch thick. Based on what was believed to be in the line at the time, the American Oil Company stated that only 10,000 gallons of oil had leaked into the river.

With the use of chemicals, plus a brisk wind and choppy seas, the oil dispersed quickly. The only areas containing any significant concentration of oil were two patches of shoreline about 200 yards long where the oil was two inches thick along the waterline. Small patches

dustrial Marine Service to do the job. The operation began on May 7, but did not include the marsh decontamination. Virginia officials felt that the clean-up action would be more detrimental to the marshes than the few scattered patches of oil that remained.

47. MAWSON METEOR TRAIL
Mawson, Antarctica   **May 9**

A white, nearly vertical atmospheric trail, probably caused by a meteor, was reported from Mawson on May 9. The trail eventually broke into a discontinuous line and was clearly visible from Mawson against the setting sun. No satellites were reported in re-entry at the time of the sighting. Four investigation parties seeking some fallen remnant were unsuccessful.

48. BURDUR EARTHQUAKE
Burdur Province, Turkey   **May 12**

A strong earthquake, registering 5.9 on the Richter scale and located about 125 miles southeast of the disastrous quake of March 28, 1970, struck the southwest region of Anatolia causing 1,200 buildings to collapse in the city of Burdur and 20 surrounding villages. The quake resulted in 57 dead and 150 officially reported injured. Seventy percent of all the buildings in Burdur and surrounding villages were damaged. On the same day, two large aftershocks of 5.2 and 5.7 Richter magnitude also occurred.

49. NORTHEASTERN MINNESOTA FOREST FIRE
Lake Superior Forest, Minnesota   May 14

A fire swept through 15,000 acres of dry woodland in the Lake Superior National Forest near Duluth, Minn. The fire was the largest since the establishment of the National Forest in 1909. Some 550 men, 17 tractors, five airplanes, and two helicopters were used to fight the blaze.

51. SHETLAND ISLANDS BIRD KILL
Shetland Islands, Great Britain   May 25

On May 28, dead seabirds began coming ashore at Grutness in southeast Shetland. By June 1, carcasses were found all along the beach. One oil slick was spotted off the island; and an even larger slick was seen about 40 miles out to sea.

Combined reports indicated that some 1,200 dead birds were counted, mostly guillemots and puffins. At least as many more birds were probably lost on inaccessible parts of the coast. And many more were killed at sea, bringing the total somewhere between 2,500 and 10,000 dead.

A sample of the oil showed it to be used fuel oil, and not crude oil as was first expected.

Northeastern Minnesota forest fire. Lake Superior Forest, Minnesota. *Photo courtesy Charles Curtis, Duluth Herald.*

Whipped by shifting winds, the fire moved five miles in eight hours, burning the entire watershed of one lake and parts of the watersheds of three other lakes. The trees and bush were extremely dry; the area had not received any rain for three weeks before the fire began.

50. MOJJO FRACTURING
Mojjo, Ethiopia   May 25

Surface cracks developed in the sandy alluvium of the rift floor at Mojjo, about 40 miles southeast of the Ethiopian capital of Addis Ababa. The cracks and holes—two to three yards deep and as much as 600 feet long—ran through villages and caused damage to some structures. No seismic activity was reported in connection with the fracturing.

52. PERSIAN GULF FISH KILL
Saudi Arabia   May 30

Great numbers of dying fish drifted ashore in the region between Jubail and Ras Tanura on the Persian Sea coastline. Most of the fish that came ashore were "black sbaitee" and "hamoor," both species of grouper, plus some angel fish. Most, too, were large mature adults weighing from one to 10 pounds, with a few weighing up to 20 pounds. A large barracuda and a large octopus were also found.

Most of the fish were still alive when stranded, coming ashore on a rising tide and pushed by a northeasterly wind. Observers reported the fish had inflated bellies. When the fish attempted to go back into the water, they were unable to remain below the surface due to their gas-

filled abdomens. The cause of the mortality was never known.

Three lots of fish from the strandings were examined in the laboratory. The specimens showed no signs of storm damage or poisoning, no damage from blasting, nor any signs of disease. The airbladders of all the fish were grossly inflated and the network of blood vessels in the dorsal wall of the airbladder was enormously distended and filled with blood, but there was little or no sign of hemorrhage into the airbladder or surrounding tissues, nor anywhere else. All the fish had empty stomachs, but were fat and otherwise in excellent condition.

In the past, similar fish kills were reported following winter storms; however, deaths in those cases were attributed to cold water. Although the stranding did occur after one of the longest sustained windstorms in recent years, the temperature had remained high.

Examination of local fish markets showed that the species involved were still being caught by the usual methods and in substantial numbers. This suggested that the kill was local or limited, or perhaps both.

53. ST. PETERSBURG RED TIDE
West Florida   June–August

A red tide, caused by the microorganism *Gymonodinium breve,* was first seen about the middle of June. Starting offshore in the Gulf of Mexico, the red tide was first noticed on the beaches from St. Petersburg to Boca Grande on Florida's west coast. The bloom fluctuated greatly. Samples taken at Tampa Bay, for example, ranged from no organisms to about 800,000 per liter of water.

In the early stages of the tide, bottom fish, such as catfish and mullet, were killed; then eels and even big jewfish weighing 100 to 300 pounds were found dead on the beach along with a variety of other species. The red tides are sporadic and may last for periods up to several months. The last one near St. Petersburg was in 1967, and prior to that, 1963.

Past blooms have usually correlated with river runoff in which various nutrients are carried downstream by land drainage. Although iron has been identified as a possible triggering mechanism for the "red tide" organism, a drought in 1971 somewhat altered the normal amount of runoff and the iron content has not been determined.

54. "WHITE FLY" PLAGUE
Spain   June–August

Approximately three million citrus trees in the Spanish provinces of Alicante and Malaga were affected by a "white fly" plague. The species was identified as *Aleurotrixus howardil* or *Aleurotrixus floccossus* of the *Homoptero Aleurodido* family.

The fly also affected plantations in the provinces of Ameria, Cadiz, Granada, Las Palmas de Gran Canaria, and Murcia.

The Spanish Ministry of Agriculture organized a massive campaign to end the plague which began in the province of Malaga in the summer of 1969, possibly brought from southern France. Before 1969, the "white fly" was very little known on the Spanish mainland, although it had been seen occasionally in the Canary Islands. Apparently the favorable weather conditions that year (high humidity, cold/warm spells, relatively little wind) caused the upsurge in the fly population.

55. ESPOLLA LAGOON RARE
CRUSTACEAN DISCOVERY
Gerona, Spain   June 3

The existence of a rarely seen crustacean, *Apus cancriformis,* was reported in the lagoon of Espolla. The species, popularly called "tortuguets," from a Catalan word meaning "small tortoise," differs from the common variety as it has two pairs of sensory antennas on each side of its head. The Espolla crustacean also has two arms on its underside ending in some rudimentary fingers covered by a membrane. The lagoon, which rarely holds very much water, was filled for five months because of unusual rains. The extraordinarily high water level apparently provided an opportunity for the crustacean to breed.

56. TASADAY MANUBO DISCOVERY
Mindanao Island, Philippines   June 7

Manuel Elizalde, Jr., a Philippine cabinet member and president of the Private Association for National Minorities (Panamin), made initial contact with the Tasaday Manubo, a food-gathering group still using stone tools and without permanent habitation. When discovered, the Tasaday had no knowledge of agriculture and had never tasted salt or sugar, or smoked tobacco. They were unaware that other societies of people existed, although they had

Tasaday Manubo discovery, Mindanao Island, Philippines. *Photo courtesy Panamin, Manila, Philippines.*

occasionally glimpsed or heard voices of hunters and trappers in the forest.

The Tasaday live in an extremely isolated mountainous area, mostly above 3,000 feet, which is covered by hundreds of square miles of rainforest. Their universe is confined to the floor of the forest which is warm, but damp and dark due to clouds which accumulate around the mountaintops. As their life is wholly restricted to this canopied rainforest, the Tasaday had no words in their language for lakes, oceans, open fields, and even constellations and phases of the moon.

Before Elizalde's first contact on June 7, 1971, a Manubo Blit hunter named Dafal had seen them about a dozen times. From Dafal, the Tasaday received a few metal objects, some brass

Tasaday Manubo discovery, Mindanao Island, Philippines. *Photo courtesy Panamin, Manila, Philippines.*

wire which they use for earrings, a few old baskets and rags of cloth, and their only musical instrument—a Jew's harp. He also taught a few men how to trap wild pigs and monkeys and

to hunt with a bow and arrow. They had no hunting or trapping devices before Dafal.

The Tasaday still use stone tools, high-angled scrapers of quartz, hafted edge-ground scrapers, and hafted stone hammers. They also use any handy river pebble for pounding. When contacted by Elizalde, they had only one metal knife—a broken and badly nicked bolo that they had received from Dafal.

They are Mongoloid people—short, with fair skin, straight or wavy hair, and high cheekbones. The Tasaday speak a Malayo-Polynesian language closely related to the Manubo subgroup of the Philippine languages found in Mindanao. A lexico-statistical study by Father Llamzon suggests that they separated from the Manubo Blit, the language to which the Tasaday language is most closely related, about 700

Tasaday Manubo discovery, Mindanao Island, Philippines. *Photo courtesy Panamin, Manila, Philippines.*

years ago. It would also appear that the Manubo Blit were formerly food gatherers who had become slash-and-burn agriculturists in the protohistoric period.

The basic diet of the Tasaday is derived from species of wild yams or Discorea. These are obtained with a digging stick pointed with a stone scraper or with the bolo that they received from Dafal. Wild fruits of trees also form an important seasonal source of food. Their major source of protein comes from large tadpoles (three to five inches in length) of the Giant Mountain Frog *(Rana magna)*, crabs, and fish, all of which are abundant in the small creeks. Dafal, the Manubo Blit hunter, also taught them how to gather and prepare the pith of wild palms which provided a new staple food for them during the past four years.

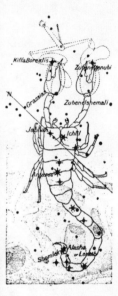

Tasaday Manubo discovery. Mindanao Island. Philippines. *Photo courtesy Panamin. Manila. Philippines.*

The 24 individuals contacted formed a local group, or *nasagbung,* which was composed of six closely related families. The family is the basic social and economic unit, although they share food gathered during the day. There is no group leader within the local group. Nor are there any formal social roles, such as that of midwife, and a woman delivers her own babies. They speak continually of the dead and they attribute all that they possess to their ancestors—the *fangul.* They say that they see their soul relatives—the *sugay*—in their dreams and state that these departed relations dwell in the crowns of trees. Since the canopied forest forms the boundaries of their universe, this dwelling place for the dead is highly significant.

Dress is provided by the leaves of *curculigo,* a tufted herb, the bark of trees, or the few rags they obtained from Dafal. They have no permanent dwellings, but live in rockshelters or between the buttress roots of giant trees, using palm fronds as covering when it rains. Most adults chew a wild betel with a mixture of lime made from land snails. A few adults had faint tattoos made with a plant dye and thorns. No ornaments were encountered except for the brass wire earrings given to them by Dafal in 1969 which replaced similar earrings made of plants.

Apparently, the Tasaday have lived in isolation for centuries, and perhaps even since prehistoric times. They are still exclusively food gatherers, doing only limited trapping and hunting with devices taught to them by Dafal during the past four years. Prior to 1967, their technology was based entirely upon stone.

57. MILAN HAILSTORM
Milan, Italy   **June 11**

The Milan region was pelted by a heavy hailstorm, with average stones as large as two centimeters in diameter and some even as big as 4.5 centimeters. Agriculture in the Lombard plain was seriously damaged by the giant hailstones, and floods were reported in a few areas of the Milan region.

Milan hailstorm. Milan. Italy. Hailstones about 2 centimeters diameter. some 4.5 centimeters. fell on Milan on June 14. 1971. *Photo courtesy Mr. Palumbo. APICFEO. Milan. Italy.*

58. LIVERPOOL NAPHTHA ESCAPE
Liverpool, England   June 12

Some 600,000 gallons of liquid naphtha, stored in tanks for use in the production of domestic gas for Liverpool, escaped when two or three valves in a surface installation beside the River Mersey mysteriously became unfastened during the night. By the time the leak was detected the next morning, much of the naphtha had soaked into the ground. Some also seeped into a sewer and onto the mudflats beside the river.

An emergency was immediately declared and a square mile of the town inhabited by 2,000 people was cordoned off for ten hours. During this time, the exposure of naked lights was prohibited and traffic was halted in the approaches to the Manchester Ship Canal. Although it was thought that there might be some danger to firemen attempting to control the escape due to the toxic action of naphtha on the blood, no serious damage other than the dislocation of activities in the neighborhood was reported. The naphtha could be smelled in Liverpool City Center, eight miles away.

59. CHILE–ARGENTINA BORDER
EARTHQUAKE
Chile   June 17

One person was killed and minor damage reported when an 80-second earthquake rocked the Chile–Argentina border. The quake, felt over an area extending some 1,000 miles, was recorded at Richter 7.0 magnitude. Government reports claimed the quake caused the collapse of old walls and the toppling of building cornices in some cities, but no other major damage.

60. WESTERN OHIO FIREBALL
Ansonia, Ohio   June 18

A very bright fireball over Ohio was also seen by observers in Indiana, Illinois, Wisconsin, and Michigan. The fireball was generally described as yellow or white and very bright. After the terminal explosions, many observers also reported seeing red objects, or "streamers."

An observer in Ansonia, Ohio, heard a very loud sonic boom three to four minutes after the fireball lit up the night sky "like noontime." The boom, he said, was followed by a rumbling sound. The fireball was generally observed from four to six seconds, and moving in a southeasterly direction.

61. VALPARAISO EARTHQUAKE
Valparaiso, Chile   July 9

A strong earthquake occurring near Valparaiso, Chile, resulted in the deaths of 90 people, hundreds of injuries, and extensive damage to buildings in that city and as far away as Santiago, 110 miles to the southwest. Landslides caused by the quake blocked highways and railroads while some 30 to 90 percent of the buildings in various towns of the area were destroyed or damaged.

In Santiago, the entire downtown area was strewn with smashed glass and broken masonry. Walls cracked and water lines burst. The earthquake reached magnitude 7.8 on the Richter scale. Across the Andes in Argentina, the shocks were felt in Buenos Aires, La Plata, and other cities. The quake knocked the needle off the seismograph in Argentina's Meteorological Observatory.

The earthquake was located very near a 230-foot dam that collapsed during a 1965 quake killing 400 people and injuring 350 others.

There were a number of aftershocks following the main tremor, but none larger than magnitude 5.0 Richter.

62. SOLOMON ISLANDS–NEW BRITAIN
EARTHQUAKES
Solomon Sea, South Pacific   July 14 and 26

Two 8.0 magnitude earthquakes occurred on July 14 and 26 about 60 miles apart in the north Solomon Sea. The aftershock patterns showed that they were the result of movement along two separate fault planes: the first extending from southernmost New Ireland to the south-southwest, and the second extending from the same point to the west-southwest.

The fault planes met or overlapped in the area south-southeast of New Ireland where the epicenter of the July 26 event was located and the fault rupture commenced, due to triggering by the initial (July 14) earthquake series.

The quakes caused tsunamis at Rabaul and along the coasts of southeastern New Britain, southern New Ireland, and western Bougainville. One or two lives were known to have been lost in the tidal waves.

For a six-month period, January to July, there had been a great concentration of large magnitude earthquakes including one of magnitude 8.1 in New Guinea, January 10; 7.8 in Chile, July 9; and 8.1 in Solomon Islands. A Similar rash of big quakes occurred between October 1963 and March 1964.

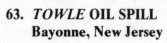

63. *TOWLE* OIL SPILL
Bayonne, New Jersey   **July 14**

The *Towle*, a civilian-manned navy cargo transport moored at Military Ocean Terminal in Bayonne, N.J., spilled 38,000 gallons of oil during fuel transfer operations. The resulting oil slick polluted beaches in the New York–New Jersey area, especially in the Gravesend Bay–Coney Island region.

No chemical dispersants were used to get rid of the oil, but vacuum pumps and flushing techniques were used successfully to retrieve about 19,000 gallons of the oil. Absorbent material was used to soak up the remainder of the slick.

64. WHITE ISLAND VOLCANIC ACTIVITY
New Zealand   **July 19**

The White Island Volcano, in the Bay of Plenty, North Island, New Zealand, erupted on July 19 ejecting a large plume of steam as well as black ash and boulders, some of which reached the sea half a mile away. An ash cloud 6,000 feet high was also produced.

The eruption formed a new vent at the south end of the 1933 crater, about 300 feet long and 150 feet wide, with an average depth of 35 feet. About six smaller vents were formed within it, emitting steam at a high velocity and at a temperature of 400 to 500 degrees centigrade. The steam from one vent was slightly yellow, indicating the presence of sulphur.

By July 20, the force of the eruption had diminished considerably, though large explosive outbursts occurred from time to time. Since the July 19 eruption the new crater continued to emit steam, but no further tephra eruptions have been reported.

65. MORONA SANTIAGO EARTHQUAKE
Ecuador   **July 27**

A strong earthquake of Richter magnitude 7.5 struck the Morona Santiago province of Ecuador near the Peru border. The quake, felt in Quito and Guayaquil, lasted for almost one minute.

66. HÄVERO METEORITE FALL
Finland   **August 2**

A round, 1,544-gram meteorite fell through the roof of a farm storehouse on the island of Havero in southwest Finland. The stone was found inside a box on the storehouse floor by the family's nine-year-old boy. Apparently, the object had passed through both the roof and the top of the box. Sonic booms accompanying the descent of the stone were heard by many people. One described it as "first like a thunderstorm, then like a jet plane, and then a big bang." The meteor fell to the storehouse in bright sunshine, thus no light phenomena were reported. (A dust cloud inside the storehouse, actually caused by the passage through the building roof, at first convinced the farm family that the stone had exploded upon impact.)

Samples of the meteorite examined at the Smithsonian Astrophysical Observatory and in other labs showed it to be a rare stony type known as a ureilite containing microscopic traces of natural diamonds.

67. HIDAKA EARTHQUAKE
Hokkaido, Japan   **August 2**

A strong earthquake of Richter magnitude 6.6, followed by five powerful aftershocks, shook Tokyo and northern Japan. Tsunami warnings were issued for the Pacific coast of Hokkaido, but later were canceled. Abnormally high sea waves and a weak tsunami at Urakawa, Hokkaido, were reported, however.

68. ROMANCHE GAP EARTHQUAKE
Mid-Atlantic Ridge   **August 5**

An unusually strong earthquake was reported near the Romanche Gap in the mid-Atlantic Ridge, near the Equator, approximately 900 miles southwest of Monrovia, Liberia. The quake was recorded at Richter magnitude 7.0.

69. LEES RIVER FISH KILL
Massachusetts   **August 5**

The Lees River fish kill involved nine species of marine fish and two species of invertebrates. In excess of one million juvenile menhaden were killed with lesser numbers of weakfish, tautog, cunner, American eel, white perch, oyster toad fish, mummichog, and silversides involved in the mortality. Approximately a half-million prawns were also killed.

Field investigations revealed the cause of the fish kill to be low dissolved oxygen. The depressed dissolved oxygen level was due to industrial and commercial discharges as well as restricted tidal circulation. Laboratory analysis of industrial effluent and seawater samples in the estuary revealed excessive nitrogen, phosphorus, and ammonia.

70. GREEN POND FISH KILL
Massachusetts   August 11

An estimated 17,000 fish were killed in Green Pond, with approximately 80 percent of the dead mummichogs. Other species in the mortality included silversides, sheepshead, minnow, and menhaden.

Field investigations showed a severe depression of dissolved oxygen as the cause of mortality, but the reason for the low oxygen levels were unknown. It was suspected, however, that pesticides played a major part in the mortality. Laboratory analysis showed that the pesticide Parathion, although not found in the water samples, was detected in the fish. However, the amounts of the pesticide were below the toxicity level.

71. MOUNT HUDSON VOLCANIC ACTIVITY
Chile   August 12-18

The Mt. Hudson volcanic eruption, first reported as a thunderstorm, sent massive clouds of ash and water vapor high into the air. Mount Hudson, located in a sparsely populated part of the Patagonian Andes, has an ice-mantled summit and is intensely dissected by glacial valleys. The caldera contains an ice field that feeds two main glaciers. The larger of the two, the Huemules glacier, is the origin of the Rio de Los Huemules.

On August 18, a column of ash and water vapor was observed rising 20,000 feet above the volcano. No lava was seen, but ash-free steam emitted from a second or third vent was interpreted as indicating the presence of a minor lava flow under the ice cover.

A severe ash fall was reported in the vicinity of the volcano, so severe that settlers and cattle were evacuated. Melting at the head of Huemules glacier mobilized ash deposits and pyroclasts, producing a lahar-like flow that advanced down the Rio de Los Huemules valley, destroying homes, farmland, and cattle. An adult and two children were reported

missing, but generally the flow was not violent. Most houses remained intact, although many filled to the ceiling with mud.

On September 18, the ash column rose 400 yards above the summit of Mt. Hudson. Also, rhythmic explosions were reported occurring every three minutes.

72. SAN CLEMENTE OIL SPILL
Southern California   August 20

The U.S. Navy tanker *Manates* spilled about 1,200 gallons of heavy fuel oil while refueling the aircraft carrier U.S.S. *Ticonderoga* off the southern California coast. The oil slick paralleled the coast near San Clemente in a broken line for about 20 miles. The marine damage resulting from the spill was considered minimal.

73. MONARCH BUTTERFLY PEAK
North America   September–November

Fred A. and Norah R. Urquhart, both of the University of Toronto, have been following the migration of the Monarch butterfly, *Danaus plexippus,* in North America since 1937. Although they were aware that there was a decided variation in population density from year to year, accurate accounts of population density were not taken until 1950. With the assistance of many hundreds of volunteer research associates, it was possible to compile statistical observational data on population fluctuations for various parts of North America.

The peak population occurred in 1971 as the Urquharts had forecasted, and "unprecedented" numbers of Monarchs were reported throughout most of North America during the summer of 1971.

A post-peak year was expected during 1971–1972 with a rapid decline again in 1973–1974.

74. TINAKULA VOLCANIC ERUPTION
Southeast Solomon Sea   September 5

A volcanic eruption on Tinakula Island, preceded by a small tidal wave, was reported September 5, characterized by flames and dense white clouds of smoke at the summit, with similar activity continuing intermittently until October 3. Red-hot blocks of rock, ejected at intervals of 30 seconds, fell onto the sides of the volcano and rolled into the sea. By September 13, slow-moving stream of red-hot lava extended down the northwest side of the volcano, at about the same location as the 1966 flow.

Approximately 160 persons living on the southeast coast were evacuated by ship on September 17 and 18, earlier evacuation attempts having been made impossible by heavy seas.

The "still-warm" flow was examined on November 21 and was found to resemble the "aa" type, consisting of poorly consolidated cinder enclosing angular fragments of rock up to a yard long. Sublimed yellow crystalline sulphur was common around cracks emitting steam and sulphur dioxide gas.

A strong earthquake on November 21 caused increased movement of the scree material, and a decrease in the number of gas-emitting vents was noticed when the volcano was revisited on November 22.

75. FUEGO VOLCANIC ACTIVITY
Guatemala   September 14–15

Tremors, hot rocks, boulders, and a rain of very fine ash characterized the Fuego Volcano eruption that began September 14 and ended suddenly the next day. During the one-day eruption, glowing molten material was thrown 3,000 feet from the crater as a huge black cloud rose high in the air. All observers agreed that the eruption was the volcano's most spectacular in 70 years.

76. BOUGAINVILLE EARTHQUAKE
British Solomon Islands   October 4

In the same general area where two 7.9 Richter magnitude quakes struck in July, another powerful tremor shook the Solomon Islands approximately 25 miles southwest of Schano, Bougainville Island, British Solomon Islands. The quake was recorded at 7.0 Richter magnitude.

77. AMOCO OFFSHORE OIL PLATFORM FIRE
Gulf of Mexico   October 16

Four of the five wells on an AMOCO oil platform, located about 125 miles southwest of New Orleans, La., caught fire on October 16 and kept burning until November 30. On October 18, there were reports of oil slick patches with a maximum width of one-quarter mile. A rainbow film extended an additional four miles, with patches of sheen extending

another five miles northwest of the platform. The oil spilled did not represent any danger to the beaches. Most of the oil was burned in the intense fire, and a pollution control boat was on the scene carrying dispersant chemicals.

78. SOUFRIÈRE VOLCANIC ACTIVITY
St. Vincent Island, West Indies   October

Great clouds of steam rising from a lake in the crater of Soufrière Volcano was a signal that the water was being displaced by lava. From November 3 to January 31, 1972, an estimated lava input of approximately 45 million cubic meters, or a daily input of 0.5 million cubic meters, flowed into the lake.

In December, an island of lava appeared above the lake surface and expanded steadily. By the end of January 1972, the island was 70 yards above the water level of the lake.

Seismic disturbances associated with the activity was slight. Apparently, this meant no significant volume of molten rock was on its way to the surface. Thus, no major eruption of the volcano was expected.

If the rate of lava emission continued, the crater could be almost completely filled by early 1973. However, the new lava is so viscous that it is unlikely that it will flow down the mountainside should it overflow the crater rim.

The eruption was the fourth recorded at Soufrière. The first three, in 1718, 1812, and 1902-03, were all extremely violent. The 1902 eruption devastated the northern third of the island and killed 1,565 people. The lake was formed in 1910.

79. LA PALMA VOLCANIC ACTIVITY
Canary Islands   October 26–November 18

On October 26, the La Palma Volcano erupted, emitting gases and tephra through three vents. Lava fountains formed two main flows that reached the sea at the southern tip of the island.

The northern vent threw lapilli and large bombs to a height of 500 to 700 yards, while the southern vent threw only bombs and scoria. Lava was emitted through the upper vent and through a point in the middle of a fracture at the southern end of the island 2,500 meters away from the vent. The activity subsided on November 18.

80. ESPÍRITU SANTO EARTHQUAKE
New Hebrides Islands   October 27

A strong earthquake occurred in the South Pacific on October 27, 30 miles south of Luganville, Espíritu Santo Island, New Hebrides. The magnitude of the tremor was recorded at 7.4 Richter.

81. UTUPUA EARTHQUAKE
Santa Cruz Island Chain   November 21

On November 21, another strong earthquake occurred in the South Pacific some 100 miles southeast of the Tinakula Volcano that erupted in September. The quake, located near Utupua Island in the Santa Cruz Chain, was recorded at 7.0 Richter magnitude. On October 27, an earthquake occurred in nearly the same area with a magnitude of 7.3 Richter.

82. *JULIANA* OIL SPILL
Niigata, Japan   November 30

At least 4,000 tons of crude oil spilled from the Liberian tanker *Juliana* after the ship ran aground and broke in two on November 30. Twelve patrol boats, two helicopters, and six fire engines were used to combat the pollution of the sea water by spreading chemical dispersants over the slick. There was also an attempt to fence in the oil slick with a 970-yard-long chain of rubber floats.

83. VILLARRICA VOLCANIC ERUPTION
Villarrica, Chile   November – December

The eruption of the Villarrica Volcano on November 29 produced lava flows that melted the mountain's ice cover, causing lahars to flow down the slopes at speeds up to 60 miles an hour. The lahars carried trees, rocks, houses, and farms, and were responsible for the deaths of at least 15 persons.

The volcano actually began its activity on October 29 with a weak gas explosion and some pyroclastic material, but it erupted more violently on November 29, ejecting lava flows and pyroclastics over its ice cap, forming a cone in its central crater.

The activity increased between December 3-20, with three major lava flows on the southwest slope. The flows caused partial melting of the cover ice and slowly formed a channel in the glacier with a depth between 20 and 40 yards. Lava flows occurred every 30 seconds in the central crater.

Activity subsided in December, although vapor emissions continued from the south fissure during January and February.

84. TELICA VOLCANIC ACTIVITY
Nicaragua   December 1

An active lava lake was observed deep within Telica's summit crater on December 1 by Dartmouth College geologists. Similar observations from the rim one year before showed no lava lake. The volcano has had numerous small ash eruptions in 1971, but none of these affected more than the immediate area of the cone.

85. LABAN ISLAND OIL SPILL
Persian Gulf   December 2

The blowout of an Iran Marine International Oil Co.'s oil well, located 80 miles off the shore of Laban Island, created a slick that covered some 350 square miles of sea surface. The oil was inches thick in some areas, but most of it merely formed a florescence on the water. The oil eventually formed an elongated slick covering an area of 800 square miles.

By December 28, the oil well had become depleted, and a big storm and high waves had broken up the slick that it generated. An estimated 100,000 barrels of oil were spilled into the sea.

86. PEACE RIVER PHOSPHATE SPILL
Florida   December 3

An extensive fish kill resulted from the breaking of a dike used to contain phosphate slime. When the dike, owned by Cities Service Co., broke on December 3, the colloidal slime poured through Widden Creek into the Peace River and spread all the way to its mouth.

In 1967 a similar kill (from another company) resulted from spillage of approximately the same amount of phosphate slime into the Peace River. During this kill, some one to two million fish, or approximately 90 percent of the total population, were destroyed. The 1967 spill occurred farther upstream, but in 1971 there was more damage to the creek. The December 3 spill affected over 70 miles of the river from Fort Meade to the mouth at Punta Gorda.

The Department of Pollution Control announced a $20 million suit for the December 3 Peace River phosphate spill. The lawsuit also asked for a temporary injunction ordering

Cities Service Co. to "take several immediate restorative and corrective actions and to cease all phosphate processing operations until the complete and immediate inspection of all their retention dams."

Some of the fresh-water fish that were affected by the pollution were bass, bluegill, bullhead, and channel catfish; gar and mudfish were also present. In addition, there was a significant effect on aquatic plants.

In the case of the 1967 spill, it took approximately 15 months for recovery of the aquatic fauna.

87. KAMCHATSKIY PROLIV EARTHQUAKE
Kamchatka, U.S.S.R.   December 15

A very strong earthquake struck the Kamchatskiy Proliv (strait) approximately 80 miles northwest of the Commander Islands. Very small tsunami waves were observed on two of the Aleutian Islands, Alaska, as a result of the quake. Also, five seawaves five yards high were recorded in Ustkamchatsk 17 minutes after the tremor. The earthquake registered Richter magnitude 7.8.

88. KILAUEA VOLCANO
Hawaii   December

Kilauea Volcano erupted from its summit and from both its east and southwest rifts during 1971. The Mauna Ulu eruption, continuous for more than two years on the upper east rift, declined and finally ended in late 1971. The decline of the Mauna Elu eruption, however, marked the onset of a period of rapid summit tumescence, which had resulted in a 10-hour eruption in the summit caldera on August 14, and then a 5-day eruption which had begun in the caldera on September 24 and subsequently migrated several kilometers down the southwest rift.

Mount Etna volcanic activity. Sicily. Italy. Lava rivers in the wood of Cubania. May 17. 1971. *Photo courtesy J. C. Tanguy. Lab. de Géomagnétisme. Paris. France.*

AFTERWORD: THE FUTURE OF THE PLANET

This encyclopedic report on the "state of the planet" for the period 1968 to 1971 might be considered by some merely an exercise in global journalism. Events are treated purely as events, regardless of whether they are profoundly catastrophic or simply bizarre. Little or no connection between the events is offered. Explanations, interpretations, or even hypotheses are few. On the surface, then, this appears only a collection of unrelated and unconnected facts baldly and blandly stated, objectively and coolly presented: the very model of a journalistic ideal perhaps unattainable in real life.

Still, the steady accumulation of facts—cold, dry, and objective, one piled upon the other in endless succession—finally produces a vivid and frightening image of a dynamic and dangerous planet. If any apparent link exists between the diverse phenomena, it is the ultimate sense of man's helplessness in the face of nature's fury and his own stupidity. If we do not die in some incredible calamity, then surely we will kill ourselves through the misuse of our environment.

But this report is not designed as a message of impending doom. This book is only one interesting and unusual by-product of a serious scientific endeavor intended to assist in saving the world rather than bearing witness to its destruction.

As clearly stated in the Introduction, the Smithsonian Center for Short-Lived Phenomena was established solely to improve the opportunities for research by receiving and disseminating information about transient natural events rapidly and accurately to the scientific community.

In the course of serving as the "scientific hotline" for reports of new earthquakes, oil spills, meteorite falls, animal migrations, and a host of other happenings, the Center has also produced some unexpected fringe benefits. One is this report.

An equally important fringe benefit has been the first documented record of the global frequency and extent of certain natural events such as volcanic eruptions. Unbelievably, until four years ago, volcanologists had no idea how many volcanoes occurred each year. The best estimate was approximately a dozen worldwide. In only its first four years of operation, the Center reported *seventy-one*!

Naturally, the mere existence of a clearinghouse for volcano reports is bound to increase the total reported each year. But that is just the point! Until now, volcanologists could not estimate the number of volcanoes, simply because they didn't know about those happening in remote and isolated corners of the world.

Why do we need to know the yearly average of volcanoes? Volcanologists suspect that eruptions follow a cyclic pattern of activity. For example, in the years between 1909 and 1912, more volcanoes were reported than in any period previous. Many scientists think we are now in another period of extreme activity. Unfortunately, the lack of complete statistical data prevents them making such a prediction with any certainty.

The watchful eyes of the Smithsonian Center may now help science sharpen the cycle. With a volcano watch maintained over several generations, combined with other detection means, the cycle eventually may show "highs" and "lows" with sufficient clarity for scientists to issue advance warnings that could prevent unnecessary death and destruction.

In addition to sharpening the volcano cycle, the Center has had another fringe benefit: its contributions to the growing awareness of the earth as a planet.

Since the advent of global mass communications, group air fares, and interplanetary spaceflights, there has been a general tendency to assume all the frontiers have been conquered, and all the secrets of the world discovered. But, after nearly a decade of seeking new goals in the stars, man is at last looking back toward his own habitat again.

The most obvious example of this scientific introspection is the Earth Resources Technology Satellite, a complex earth-orbiting observational system that measures a variety of phenomena on the surface, in the interior, and in the atmosphere of the earth. The Center for Short-lived Phenomena is one of the many "experimenters" aboard the ERTS launched by the National Aeronautics and Space Administration in 1972.

The Center provides the ERTS with "ground integrity," that is, earth-based observational back-up to those events seen from space. In reality, the Center and satellite work together, one alerting the other to ongoing events. Typically, a Center correspondent might report a large-scale event such as a tidal wave or an insect plague. The Center would alert the ground-controllers of the satellite that such an event was in progress, with the accurate ground coordinates and approximate area of coverage. The satellite, in turn, would be programmed to begin surveillance and recording observations of the event. In the case of truly massive events, such as forest fires or floods, the satellite may provide overall information on the intensity and extent of the damage as well as pinpointing other potential danger areas.

The Center contributes to still another, more subtle, form of environmental awareness through its enhancement of science curriculums in schools around the world. Center announcement cards and event reports are being used as effective teaching tools in hundreds of college and secondary school classrooms. Arriving with the urgency and impact of airmailed bulletins, these daily and weekly reports recounting red tides, meteorite falls, discoveries of tribes, animal infestations, or land fracturings help turn classrooms into lively science news centers. The receipt of reports about actual ongoing geophysical and biological events adds a sense of relevancy and immediacy to what could otherwise be a dry and static presentation of earth science and general biology. The idea of listening in on a global hotline captures the imagination of young people everywhere.

The Center is also exploring the means to make schools—more specifically, individual teachers and students—even more active participants in its operations. Ideally, the Smithsonian would like to see students around the world serve as Center correspondents. In return for receiving news of the changing conditions elsewhere, young people organized into loose regional groups would report any major events in their own locales. Ultimately, the Center would like to see the planet blanketed with a youthful reporting network responsive to scientific needs.

While the idea of an "environmental kiddie corps" may never be realized, the broader concept of worldwide monitoring systems seems an imminent reality. Although concern for the planet is not new, man's modern ability to observe the total world on a continuing basis is unprecedented. For the first time in history, man how has the tools to collect, collate, and evaluate the multitude of facts that can help him understand the forces—both natural and man-made—shaping the destiny of his planet.

Just such a worldwide environmental monitoring system was proposed at the first United Nations Conference on the Human Environment held in Stockholm in June 1972. The suggested system, dubbed "Earthwatch" by its designers, received almost unanimous approval (with the exception of how to pay for it) from the representative nations at the conference in Sweden.

The Earthwatch system was developed along guidelines laid down by an international group of scientists, including members of the Smithsonian Institution staff. The total system would feature:

1. Ten baseline pollution monitoring stations located in deserts, tundra, and high mountains where there is little local pollution. These stations, thus, would both provide standards against which to measure air pollution levels and watch for any long-term and subtle changes in the global atmosphere.

2. Over 100 local and regional air-quality stations to measure atmospheric conditions over specific urban areas as well as over larger areas sharing similar pollution problems.

3. Oceanographic stations, including submarine sites, to produce the baseline data necessary for long-range studies of aquatic and marine ecosystems and to monitor any changes in local water environments.

4. International research centers and biological stations to monitor soil changes and long-term reductions or changes in animal and plant life.

5. An international system for monitoring food contamination by chemicals, pollutants, and pesticides.

6. A central clearinghouse to coordinate the efforts of existing systems and networks.

Obviously, the message from Stockholm was explicit. To save our world, we must have more information. Precise and complete data on both the general and specific conditions of our oceans and atmospheres and lands are necessary before we can make any long-range environmental decisions for the future. We need a constant flow of facts and figures from the world's various geographical and climatic zones, from country and city, desert and jungle, mountain and plain.

The baseline levels of pollution must be established, the cyclic pattern of natural events recorded, and the absolute limits of Earth's endurance determined. Only then can man understand his planet and make plans for preserving it.

Earthwatch will be the means to this end. And in its establishment and operation, Earthwatch will owe much to the pioneering efforts of the Smithsonian Institution and its Center for Short-Lived Phenomena. Indeed, the Center's most valuable and lasting fringe benefit may be the example it has set for international cooperation.

As the prototype of the monitoring systems that must surely follow, the Center has shown that worldwide reporting networks, even when composed of loosely bound volunteers, can contribute to an overview of the planet. In this context, too, the eclectic and apparently haphazard collection of "events" takes on a new significance: those reports are vital desposits in the world's own natural data bank.

No doubt the Smithsonian Center will eventually become an integrated part of the international monitoring network: one more cog in a vast organization. However, the Smithsonian's contribution will remain unique, for its transmission lines will continue to carry news of strange deviations from the norm.

Let other systems monitor the standards and the baselines, the gradua! ebb and flow of planetary rhythm. The Center for Short-Lived Phenomena will persist in its role as global wire service: providing its existential record of those arbitrary, pernicious, unpredictable, unusual, unforeseeable, and often unexplainable cataclysmic events that quicken the pulse of the planet and the hearts of its inhabitants.

Bookstore.

APPENDIX A
SELECTION CRITERIA FOR EVENTS REPORTED BY THE CENTER

Earth Sciences Events: Earthquakes greater than magnitude 7.0 or earthquakes occurring in unusual areas or those creating exceptional interest. Crustal movements, faulting and fissuring, major land movements, and landslides.

Volcanic eruptions, submarine eruptions, the birth of new islands, island eruptions, the disappearance of islands, caldera collapses, fissure extrusions, *nuées ardentes*, and major mudflows.

Earthquakes under the sea floor greater than magnitude 7.0 or having a considerable effect on the marine geophysical environment. Island earthquakes, tsunamis, sea surges, and severe storm erosion.

Polar and subpolar events, formation of ice islands, unusual sea ice break-ups, surging glaciers, and sudden release of glacier-dammed water.

Biological Sciences Events: Sudden changes in biological and ecological systems, invasion and colonization of new land by animals and plants, rare rapid migrations, unusually abundant reproduction or death of vegetation, establishment or re-establishment of flora and fauna.

Severe climatic changes affecting ecosystems, ecological aftereffects of short-term human intrusion into an area previously unvisited by man, and potentially imminent species extinction.

Sudden changes to marine and aquatic environment, oil pollution, unusual occurrences of marine vegetation, marine bioluminescence, red tides, plankton blooms, and fish kills.

Fires that have a major ecological impact on animals and flora; those that have a major environmental impact and that cause major devastation.

Astrophysical Events: Large fireball events, meteorite falls, and crater-producing impacts. Transient lunar events: obscurations on lunar surface, brightenings, lunar volcanic activity, moonquakes, and meteorite impacts recorded by implaced lunar seismometer.

Urgent Archaeological Events: Discovery of archaeological sites threatened with imminent destruction.

Urgent Anthropological Events: Newly discovered tribes; rapid changes in human ecological systems; short-lived acculturation; dying languages, customs, and people; and major human migrations.

APPENDIX B
SUMMARY OF SHORT-LIVED PHENOMENA
REPORTED 1968-1971

EVENT	1968	1969	1970	1971	TOTAL
Volcanic eruptions	12	18	22	19	71
Major earthquakes	18	29	19	20	86
Major oil spills	6	12	16	17	51
Major fireball events	12	14	8	6	40
Meteorite falls	1	5	5	4	15
Major animal mortalities	5	20	12	7	44
Major animal migrations, population fluctuations, infestations, colonizations, outbreaks, invasions, and plagues	3	9	8	10	30
Major pollution/ecological events	3	8	6	3	20
Major landslides, landslips, avalanches	1	7	2	2	12
Major storm surges, tidal waves, and floods	1	4	4	2	11
Unusual geological events	3	5	5	2	15
Major vegetation events, red tides, forest fires	2	4	3	2	11
Urgent anthropological events	0	2	1	1	4
Urgent archaeological events	1	3	0	0	4
Other events	2	5	2	4	13
Total events	70	145	113	99	427

Scientific teams investigated 336 of the 427 events reported by the Center in the years 1968-1971.

APPENDIX C
CORRESPONDENTS OF THE CENTER
FOR SHORT-LIVED PHENOMENA

To establish an effective global reporting network for short-lived events, the Center invited scientists in many disciplines and from many countries of the world to join the Center as correspondents. The initial response to the establishment of the Center by the international scientific community was overwhelming. Within just a few months the Center was routinely reporting virtually every major volcanic eruption, earthquake, meteorite fall, and oil spill that occurred in the world. Within a year after its inception, the Center had 780 registered scientific correspondents in 74 countries and had reported 70 short-lived events.

By January 1972, the Center had developed a reporting network of over 2,700 correspondents in 143 countries and territories on all the continents and oceans of the world and had communicated information on 427 short-lived events.

Correspondents are scientists, scientific institutions, or field stations that cooperate with the Center by reporting events that occur in their respective areas, sometimes traveling to the events to make follow-up reports, and occasionally providing assistance to research teams sent to investigate the events. In return, they receive notification of short-lived events occurring around the world. The following table lists the number of correspondents by country:

Country	No.	Country	No.
Afghanistan	2	El Salvador	2
Albania	4	England	121
Algeria	5	Ethiopia	6
Angola	4	Fiji Islands	6
Argentina	20	Finland	14
Australia	77	France	73
Austria	12	French Guiana	1
Azores	1	French Polynesia	6
Belgium	11	Gabon	4
Bermuda	4	Gambia	1
Bolivia	9	Germany	86
Botswana	6	Ghana	8
Brazil	49	Gibraltar	2
Brunei	4	Greece	6
Bulgaria	11	Guam	1
Burundi	3	Guatemala	4
Cameroun	5	Guinea	3
Canada	99	Guyana	7
Central African Rep.	3	Haiti	3
Ceylon	5	Honduras	6
Chad	1	Hungary	16
Chile	26	Iceland	10
China	19	India	41
Colombia	28	Indonesia	19
Congo	18	Iran	11
Costa Rica	9	Iraq	6
Cuba	5	Ireland	10
Cyprus	6	Israel	30
Czechoslovakia	21	Italy	36
Dahomey	1	Ivory Coast	6
Denmark	13	Jamaica	7
Dominican Rep.	1	Japan	78
Ecuador	10	Jordan	3
Egypt	11	Kenya	9

Korea	3	Rwanda	6
Kuwait	2	Samoan Islands	1
Lebanon	5	Saudi Arabia	8
Lesotho	1	Scotland	19
Liberia	8	Senegal	5
Libya	3	Sierra Leone	5
Madagascar	3	Singapore	5
Malawi	8	Solomon Islands	2
Malaysia	5	Somali Rep.	2
Mali	6	South Africa	21
Mauritania	1	Spain	19
Mauritius	7	Sudan	5
Mexico	36	Surinam	1
Monaco	5	Swaziland	3
Morocco	7	Sweden	25
Mozambique	4	Switzerland	54
Nepal	3	Syria	4
Netherlands	41	Taiwan	8
New Caledonia	6	Tanzania	9
New Guinea	9	Thailand	9
New Hebrides	1	Togo	4
New Zealand	42	Tunisia	4
Nicaragua	6	Turkey	7
Nigeria	5	Uganda	4
Norway	11	U.S.A.	833
Pakistan	9	U.S.S.R.	53
Panama	11	Upper Volta	4
Paraguay	3	Uruguay	7
Peru	20	Venezuela	13
Philippines	12	Vietnam	5
Poland	13	Virgin Islands	5
Portugal	16	West Indies	14
Puerto Rico	8	Yemen	1
Rhodesia	11	Yugoslavia	7
Rumania	10	Zambia	10

U.S. embassies, consulates, consulates general, missions, and offices through the world: 76

TOTAL: 2,744

APPENDIX D
SERVICES OF THE CENTER FOR SHORT-LIVED PHENOMENA

Event Telephone/Telegram Notification Service: The Center's "Telegram Event Notification Service" provides an immediate alert for those scientists, agencies, and institutions with fast-response capabilities. Notification is made as soon as possible after the Center receives an initial event report. In addition, telephone and telegram follow-up reports are made following any significant changes in the status of continuing events.

Event Notification and Information Cards: The Center notifies a second level of scientists of the occurrence of short-lived events through the issuance of "Event Notification and Information Cards." These cards are issued within 24 hours of receipt of event information by the Center. Notification cards contain initial information and data for an event. Information cards carry up-to-date data subsequently obtained pertaining to the event.

Both services are available to schools, industry, and the general public at modest subscription rates.

The Telephone/Telegram Service may be of most interest to those companies and government agencies specializing in environmental research or to the news media. Rates are based on an average charge per report, and range from a minimum of $25 for 5 reports to $300 for 90 reports.

Teachers, students, amateur science groups, schools and colleges, libraries and documentation centers, may find the Event Card Service more economical as well as more appropriate to their needs. Weekly summary and continuing status reports of all events covered by the Center will be mailed to subscribers for an annual fee of $15.

For more information on fees (which are subject to change), as well as other services and publications, write to Smithsonian Institution, Center for Short-Lived Phenomena, 60 Garden Street, Cambridge, Mass. 02138.

APPENDIX E
REPORTS ISSUED BY THE
CENTER FOR SHORT-LIVED PHENOMENA

1971

1. Center for Short-Lived Phenomena, *Scientific Information and Communications Accomplishments in 1970*, 22 February 1971.

2. Citron, Robert, *Notes on the Development of a United Nations Natural Disaster Program Prepared for the Office of Science and Technology of the United Nations*, March 1971.

3. Qudrat-i-Khuda, Muhammad, *The Bay of Bengal Storm Surge 12-13 November 1970*, 1 March 1971.

4. Viramonte, José G., *The 1971 Eruption of Cerro Negro Volcano, Nicaragua*, 15 April 1971.

5. Center for Short-Lived Phenomena, *Annual Report, 1970*, May 1971.

6. Citron, Robert, *The Establishment of an International Environmental Monitoring Program— A Plan for Action*. Prepared for the United Nations Conference on the Human Environment, Stockholm, Sweden, June 1972, May 1971.

7. Ministry of Industrialization and Mines, Management of Mines and Geology, Nouakchott, Mauritania, *The Kiffa Meteorite Fall of 23 October 1970*, May 1971.

8. Rittman, Alfredo, *The Mt. Etna Volcanic Eruption of 1971. The Volcanology Institute of the University of Catania and the International Institute of Volcanology of the National Research Council. Event Chronology. April 1971*, 3 May 1971.

9. Tazieff, H., and Le Guern, F. *Tectonic Nature and the Mechanism of Etna's Eruption of April-May 1971*, 4 June 1971.

10. Redhead, R. E., *Armyworm* (Spodoptera exempta) *Predation by Yellownecked Spurfowl* (Pternistis leucoscepus) *in the Longido Game Controlled Area*, July 1971.

11. Schoen, Ivan L., *Report of the Emergency Trip Made by the West Indies Mission to the Akurio Indians June 1971*, July 1971.

12. Elizalde, Manuel, Jr., *The Tasaday Forest People. A Data Paper on a Newly Discovered Food Gathering and Stone Tool Using Manubo Group in the Mountains of South Cotabato, Mindanao, Philippines*, July 1971.

13. Citron, Robert, *Outline for a Feasibility Study for the Establishment of an International Natural Disaster Warning System*. Prepared for the Office of Science and Technology of the United Nations, July 1971.

14. Rittman, Alfredo, *The Mt. Etna Volcanic Eruption of 1971. The Volcanology Institute of the University of Catania and the International Institute of Volcanology of the National Research Council. Event Chronology April 23-June 14, 1971* (Continued from 3 May 1971 report), 9 July 1971.

15. Smithsonian Institution, *Natural Disaster Research Centers and Warning Systems: A Preliminary Survey*, July 1971.

16. Cameron, Winifred Sawtell, *Comparative Analyses of Observations of Lunar Transient Phenomena*, October 1971.

17. Tomblin, J. F., and Sigurdsson, H., *The Soufrière Volcanic Eruption, St. Vincent Island, Caribbean Sea. Event Chronology 1 November 15-December 1971*, 20 December 1971.

1970

1. Citron, Robert, *An Approach to the Organization of the UNESCO "Man and the Biosphere Program,"* 5 January 1970.

2. Citron, Robert, *MABNET: The Establishment of a Global Network of Eco-stations for the "Man and the Biosphere" Program*, 2 February 1970.

3. Smithsonian Institution Center for Short-Lived Phenomena, *Annual Report, 1969*, 15 February 1970.

4. Davies, R. A., *The 1970 Eruption of Mt. Ulawun, New Britain*, 15 May 1970.

5. *Transient Lunar Phenomena Reports from the LION During the Apollo 13 Mission*, 22 May 1970.

6. Thorarinsson, Sigurdur, *The "Hekla Fires": A Preliminary Report of the 1970 Mt. Hekla Volcanic Eruption*, 15 June 1970.

7. Enrique, Gajardo W., *Preliminary Report on Damages Caused by the Peru Earthquake of May 31, 1970*, 23 June 1970.

8. Citron, Robert, *International Environmental Monitoring Programs*, 1 July 1970 (Revised 10 August 1970)

9. Citron, Robert, *Monitoring the Planet*, Background paper for the M.I.T. Summer Study on "Critical Environmental Problems," 1-30 July 1970, 2 July 1970.

10. Citron, Robert, *A Proposed International Monitoring Program for Critical Global Environmental Problems*, 26 July 19 0.

11. Bleahu, Marcian, and Constantinescu, Liviu, *The Floods of Southeastern Europe*, 27 July 1970.

12. Citron, Robert, *A Proposed Organization Plan for an International Monitoring Program for Critical Global Environmental Problems*, 29 July 1970.

13. Hedervari, Peter, *A Detailed Account of the Landslide Near Dunaföldvár, Hungary, 15 September 1970*, 21 October 1970.

14. *A Directory of National and International Environmental Monitoring Activities*, October 1970.

15. Yohner, Art, *Contact with a Group of Akurio Indians of Surinam*, 5 November 1970.

16. Citron, Robert, *The Establishment of an International Environmental Monitoring Program— A Plan for Action*, 6 November 1970.

1969

1. *Report on Center Communications Support for Transient Lunar Phenomena During the Apollo 8 Mission*, Smithsonian Institution Center for Short-Lived Phenomena, 5 January 1969.

2. Flyger, Vagn, *The 1968 Squirrel "Migration" in the Eastern United States*, 20 January 1969.

3. DiScala, Lanfranco, and Viramonte, J.G., *Preliminary Report on the 1968 Eruption of Cerro Negro Volcano, Nicaragua*, 20 January 1969.

4. Smithsonian Institution Center for Short-Lived Phenomena, *Annual Progress Report, 1968*, 1 February 1969.

5. Schoen, Ivan L., *Report on the Second Contact with the Akurio (Wama) Stone Axe Tribe, Surinam, September 1968*, 4 February 1969.

6. Hadikusumo, Djajadi, *Preliminary Report on the Mt. Merapi Volcanic Eruption, Indonesia, 7 January 1969*, 10 February 1969.

7. Crocker, William H., and Schoen, Ivan L., *Notes on the Discovery of the Akurio Stone Axe Tribe, Surinam*, 20 February 1969.

8. Citron, Robert, "Need for a Global Network for Environmental Monitoring," *Environmental Surveillance Working Group Report*, National Academy of Sciences, 24 April 1969.

9. *Operations Plan for the Transient Lunar Phenomena Observing Program During the Apollo 10 Mission,* 28 June 1969.

10. *Communications Support for Transient Lunar Phenomena During the Apollo 10 Mission,* 28 May 1969.

11. *Transient Lunar Phenomena Reports from the Lunar International Observers Network During the Apollo Mission,* 28 June 1969.

12. *Operations Plan and Observing Schedule for the Transient Lunar Phenomena Observing Program During the Apollo 11 Lunar Mission,* 27 June 1969.

13. *Communications Support for Transient Lunar Phenomena During the Apollo 11 Lunar Mission,* 25 July 1969.

14. *Transient Lunar Phenomena Reports from the Lunar International Observers Network During the Apollo 11 Mission,* 25 August 1969.

15. *Operations Plan and Observing Schedule for the Transient Lunar Phenomena Observing Program During the Apollo 12 Mission,* 17 October 1969.

16. *Smithsonian Institution Center For Short-Lived Phenomena Scientific Information and Communications Accomplishments in 1969,* 1 November 1969.

17. *Communications Support for Transient Lunar Phenomena During the Apollo 12 Mission,* 28 November 1969.

18. *Transient Lunar Phenomena Reports from the Lunar International Observers Network During the Apollo 11 Mission,* 19 December 1969.

1968

1. "The Submarine Volcanic Eruption and Formation of a New Island at Metis Shoal, Tonga Island," *Status Report,* No. 1, 15 February 1968.

2. Melson, William G., and Lundquist, Charles, "The 1967-68 Eruption at Metis Shoal, Tonga Islands," *Event Report,* 1 March 1968.

3. "The Submarine Volcanic Eruption and Formation of a Temporary Island at Metis Shoal, Tonga Island," *Event Report,* 15 March 1968.

4. "Smithsonian Institution Center for Short-Lived Phenomena, *Progress Report* No. 1, 1 January-31 March 1968," 1 April 1968.

5. "Mt. Mayon Volcanic Eruption, Philippine Islands," *Event Status Report,* No. 1, 20-24 April 1968.

6. "Mt. Mayon Volcanic Eruption, Philippine Islands," *Event Status Report,* No. 2, 25 April-10 May 1968, 14 May 1968.

7. "Mt. Mayon Volcanic Eruption, Philippine Islands," *Event Report,* 10 June 1968.

8. Moore, James G., and Melson, William G., "*Nuées Ardentes* of the 1968 Eruption of Mayon Volcano, Philippines," *Preprint,* 13 June 1968.

9. "The Submarine Volcanic Eruption and Formation of a Temporary Island at Metis Shoal, Tonga Islands," *Event Report* (Revised 15 June 1968).

10. "The Fernandina Caldera Collapse, Galapagos Islands,"*Preliminary Event Report,* 1 July 1968.

11. "The *World Glory* Spill, South Africa," *Event Report,* 1 July 1968.

12. Melson, William G., "The 1968 Eruption of Mayon Volcano, Philippines: Petrology of the *Nuées Ardentes* and Associated Extrusives," *Preprint,* 8 July 1968.

13. "Smithsonian Institution Center for Short-Lived Phenomena, *Progress Report,* No. 2, 1 April-30 June 1968," 22 July 1968.

14. "The Fernandina Caldera Collapse, Galapagos Islands, 11-26 June 1968," *Event Report* (Revised 10 August 1968).

15. "Mt. Arenal Volcanic Eruption, Costa Rica, 29 July -14 August 1968," *Event Report,* 7 October 1968.

16. "Appalachian Squirrel Migration, Appalachian Mountain Areas, September 10-28, 1968," *Preliminary Event Report,* 7 October 1968.

17. Simkin, Tom, "Mt. Arenal Volcanic Eruption," *Event Chronology,* 31 July–2 August 1968, 11 October 1968.

18. "Appalachian Squirrel Migration, Eastern United States," *Event Report,* 4 November 1968.

19. Melson, William G., and Saenz R., Rodrigo, "The 1968 Eruption of Volcan Arenal, Costa Rica, Preliminary Summary of Field and Laboratory Studies," *Preprint,* 1 November 1968.

20. Simkin, Tom, and Howard, Keith, "The 1968 Fernandina Caldera Collapse, Galapagos Islands," *Preprint*, 16 December 1968.

21. "Cerro Negro Volcanic Eruption, Nicaragua," *Event Report,* 17 December 1968.

APPENDIX F
SPECIAL SUBSCRIPTION RATE FOR READERS OF
THE PULSE OF THE PLANET

Please enter my subscription to the Center's event notification and information card service for a period of one year.

I understand that the cost of the subscription is *$15.00* per year for any or all categories of events issued by the Center.

Event cards will be mailed at the end of each week.

Name _____

Address _____

_____

City, State, Zip Code _____

Country _____

Please check the category (ies) desired:

___ Earth Science Events

___ Biological Science Events

___ Astrophysical Events

___ Urgent Anthropological Events

___ All Event Cards (All of the above)

Payment Enclosed ___   Please Bill Me ___

Please make check payable to "Smithsonian Institution, *CSLP*"

Send to: Smithsonian Institution Center
                    For Short-Lived Phenomena
              60 Garden Street
              Cambridge, Mass. 02138
              ATTN: Mrs. Eileen C. Cavanaugh

128

APPENDIX G
EVENT REPORT FORM

The Smithsonian Institution Center for Short-Lived Phenomena welcomes—indeed, seeks—the assistance of the public in reporting events of scientific interest that might otherwise go uninvestigated. Science classes, ecology action groups or concerned individuals wishing to report urgent environmental problems or unusual natural events may use the general EVENT REPORT FORM below. If necessary, the Center will contact the correspondent for more detailed information. Send the form Air Mail Special Delivery to:

> Mr. David Squires, Operations Officer
> Smithsonian Institution Center for
>   Short-Lived Phenomena
> 60 Garden Street
> Cambridge, Mass.   02138

TYPE OF EVENT: _____

                                  (Volcanic eruption, oil spill, fireball, archaeological find, etc.)

DATE(S) OF OCCURRENCE: _____ **TIME:**_____

LOCATION:_____

                       (City, State, Country, name of body of water, island, etc.)

LATITUDE AND LONGITUDE:_____

DESCRIPTION OF EVENT: *Please be as thorough as possible. Use additional paper if necessary. (Every different type of event requires specific questions to be asked. For example, when an oil spill occurs, it is important to know the type and quantity of oil spilled, the area affected, etc. For a fireball, the direction of travel, brightness, length of time observed, and reports of sonic phenomena are important. Please list all details; even those that may appear to be minor can be very helpful.)*

INVESTIGATORS: *List names, addresses, and telephone numbers of any persons, groups, agencies, etc. that are currently investigating the event, or are preparing to do so.*

REPORTER'S NAME:_____ **TELEPHONE:**_____

ADDRESS: _____

_____ **DATE FORM MAILED:**_____